Unruly Complexity

Unruly
Complexity

Ecology, Interpretation, Engagement

Peter J. Taylor

The University of Chicago Press
Chicago and London

Peter J. Taylor is associate professor and director of the Program in Science, Technology, and Values and the Critical and Creative Thinking Program at the University of Massachusetts, Boston, where he teaches environmental studies, science studies, and reflective practice.

The University of Chicago Press, Chicago 60637
The University of Chicago Press, Ltd., London
© 2005 by The University of Chicago
All rights reserved. Published 2005
Printed in the United States of America
14 13 12 11 10 09 08 07 06 05   1 2 3 4 5

ISBN: 0-226-79035-5 (cloth)
ISBN: 0-226-79036-3 (paper)

Library of Congress Cataloging-in-Publication Data

Taylor, Peter J., 1955–
    Unruly complexity : ecology, interpretation, engagement / Peter J. Taylor.
        p. cm.
    Includes bibliographical references and index.
    ISBN 0-226-79035-5 (cloth : alk. paper) — ISBN 0-226-79036-3 (pbk. : alk. paper)
    1. Ecology—Philosophy. 2. Ecology—Simulation Methods. 3. Science—Social
    aspects. I. Title.

    QH541.15.S5T39 2005
    577′.01′1—dc22   2005002891

⊗ The paper used in this publication meets the minimum requirements of the American National Standard for Information Sciences—Permanence of Paper for Printed Library Materials, ANSI Z39.48-1992.

Contents

Acknowledgments

This book had a long gestation. Susan Abrams at the University of Chicago Press kept faith that a manuscript would appear on her desk, but by the time it did, her declining health left her unable to complete the editorial "midwoofery." Christie Henry stepped in and guided me through review and revision so the book would appear in the world. Looking back, I can see that, starting well before the book was ever conceived, many people were influencing its eventual formation.

Don Byth, Michael Deakin, Noirin Malone, Alan Roberts, Geoff Sharp, and Geoff Watterson guided my inquiries during the 1970s in Australia and stretched my intellectual and political development. After I decided to work for a time overseas and landed in England, James Simpson introduced me to Raymond Williams's essay "Ideas of Nature," which complicated my thinking about drawing social lessons from scientific accounts of nature. The *Radical Science Journal* meetings provided a forum for examining not only science as social relations, but also the psychological depths of human natures; I value the link Les Levidow continues to provide to these discussions.

When I moved to the United States in 1980, Dick Lewontin and Dick Levins provided a model of treating one's scientific work as a political project (which I discuss in the prologue). Officially I became a graduate student in ecology and evolutionary biology, but I also found three other homes: Harvard's History of Science Department, where John Beatty and Everett Mendelsohn encouraged my interest in history, philosophy, and sociology of biology; "The Pumping Station," a wide-ranging discussion group hosted by Iain and Gill Boal, which helped overcome the isolation of critical intellectual work (as Iain continues to do in Berkeley); and the Environmental Sciences Division at the Oak Ridge National Laboratory, where Don DeAngelis and Mac Post generously supported my dissertation research in theoretical ecology. I hope Don and Mac can see that I have continued to build on that collaboration even though I did not stay within the boundaries of ecological science.

My disposition to cross disciplinary boundaries was encouraged by the biennial summer meetings of what came to be called the International Society for the History, Philosophy, and Social Studies of Biology (ISHPSSB)—thanks to Michael Bradie for first enticing me to attend as I finished my doctorate in 1985. A Mellon Fellowship in Science, Technology, and Society at the Massachusetts Institute of Technology, along with my first full-time teaching position in "Science, Technology, and Power" at the New School for Social Research, made an academic career that combined conventional and interpretive inquiries about science seem credible. During a Ciriacy-Wantrup Fellowship at the University of California, Berkeley, Michael Watts, Matt Turner, Jesse Ribot, and Raúl García-Barrios showed me how geographers and anthropologists were already addressing problematic boundaries and intersecting processes in ways that illuminated the issues I wanted to highlight in ecology. Yrjö Haila arranged an extended visit to the University of Helsinki for me to give lectures and run a workshop (discussed in chapter 5) and to explore our joint interests in ecology, philosophy, and interpretation of science. With Yrjö, as with Iain Boal and Raúl García-Barrios, after we had met as visitors to the United States and shared perspectives as outsiders, we became friends and long-term colleagues in traversing boundaries.

As I began to delve into historical and social studies of science, I became acquainted with four other scholars who continue to serve as models for intellectual persistence and generous support of colleagues. Susan Oyama exposes and probes nooks and crannies that others had overlooked, ever ready to express reservations about whether she has fully understood and communicated the problems she is chewing on. Donna Haraway, a weaver of connections, was able to name my project before I could—in 1993 she commented: "Taylor asks a deceptively simple question: How may complexity be represented, and what is the relation between the representation and its implicated practices of intervention?"

Sharon Kingsland understands the consistent work needed to compose solid and lucid history while she makes the space to give serious attention to the work of others. Diane Paul finds ways to frame and pursue important new questions in well-covered terrain.

The book project had taken form by 1990 when I moved to Cornell University, where my primary responsibility of teaching in the Biology and Society program allowed me to pursue my interest in feeding interpretation of science back to influence science's ongoing development. The book, however, was only a strand of a larger project. Collaborations, programs, and other activities were needed to promote collegial interactions across disciplines and open up space for reflexive, critical scientific practice. I had a lot to learn about fostering change, securing the support of audiences, colleagues, and institutions, and developing the personal resources to find new opportunities within difficulties. My allies in this larger project have been many.

Cornell students Derek Hall, Dieter Hollstein, Chris London, Bill Lynch, Corinne McCamey, Gwen Mills, Joanna Moresky, and Chris Scott helped sharpen my efforts to bring social interpretation of science into the teaching of life and environmental science students. Cornell faculty members Fred Buttel, Davydd Greenwood, Peter Schwarz, and Zellman Warhaft were supportive of this hybrid endeavor in science and interpretation. Raúl García-Barrios arranged for me to be a visiting professor at Centro de Investigación y Docencia Económicas, México, for two summers, where I first collected my concerns under the umbrella term "unruly complexity." When it became clear that I had to look beyond Cornell to pursue my interests, a wider set of colleagues and counselors helped me reevaluate my experiences and move in positive, new directions: Dick Burian, Tom Carroll, Adele Clarke, Julie DeSherbinin, Giovanna DiChiro, Victor Donnay, Gary Downey, Paul Edwards, Peter Elbow, Tony Gaenslen, Robert Heasley, Rachel Joffe, Wendy Kohli, Kathy Lilley, Judy Long, Mary K. Redmond, Elizabeth Reed, Mary Renda, Doug Shire, Dan Tapper, and Gail Walker. Mac Brown and Duncan Holmes led me into the world of facilitation of participatory processes. Mac, together with ISHPSSBers John Jungck and Steve Fifield, encouraged my more systematic attention to innovation in teaching. Chuck Dyke nudged me to shift my language from intervention to engagement. Yrjö Haila's "How Does Nature Speak" workshops gave me room to play as a facilitator and participant with the interaction of complexity and (apparent) simplicity. Brenda Dervin and Barbara Love gave me models of writing yourself into a book.

A valuable two-year period of exploration, discussion, reflection, and writing was made possible by a Rockefeller Fellowship at the Rutgers

University Center for Critical Analysis of Contemporary Culture and the Lang Professorship for Social Change in the Biology Department at Swarthmore College—thanks to Ruthie Gilmore and Scott Gilbert for opening up these opportunities. They prepared me well for my current position coordinating the Graduate Program in Critical and Creative Thinking at the University of Massachusetts Boston, where I continue to learn from working with experienced educators and other midcareer professionals about ways to promote reflective practice. This is mostly material for classes, workshops, and future writing, but let me acknowledge the support of my CCT colleagues—Larry Blum, Nina Greenwald, Arthur Millman, Steve Schwartz, and Carol Smith—as well as Peter Kiang and Diane Paul, in holding onto space for my interdisciplinary—sometimes un-disciplinary—work.

A range of publications have been worked into the chapters of this book. I acknowledge the permission of the following publishers and of my coauthors, Ann Blum and Yrjö Haila, to draw on these previously published works: *Oikos* for chapter 1, section A (Taylor 1988c); *Oikos* and *Biology & Philosophy* for chapter 2 (Taylor 1989b, 2000c); *Journal of the History of Biology* for chapter 3, section A (Taylor 1988a); *Biology & Philosophy* for chapter 3, section B (Taylor and Blum 1991a, 1991b); *Perspectives on Science* for chapter 4, section A (Taylor 1995a); Princeton University Press for chapter 4, section B, and chapter 5, section A (Taylor 1992b); the Ecological Society of America and Philosophy of Science Association for chapter 5, section B (Taylor and Haila 1989; Taylor 1990); Oxford University Press for chapter 5, section C (Taylor 1999b); *Science as Culture* (Taylor 1998b), Rowman and Littlefield (Taylor 2002a), for chapter 6; and University of Minnesota Press for the epilogue, section A (Taylor 2004).

Many colleagues commented on these papers in draft or in public presentations, and a number of students provided research assistance. Their contributions have been acknowledged in the original publications, but let me single out some for special mention here. H. T. Odum (chapter 3) and Tony Picardi (chapter 4) gave their time and attention to my interviews and provided access to otherwise unavailable material, as did Michael Glantz of the National Center for Atmospheric Research (chapter 4, section B, and chapter 5, section A). Greg Tewksbury suggested mapping to me (chapter 5, section B). Raúl García-Barrios and Remko Vonk allowed me to render their work in my own words and diagram (chapter 5, section C, and the epilogue, section A), and Louise Buck's comments refined my account of her collaboration with Vonk. Once the manuscript was completed, the editorial comments of Glenn Adelson, Ann Blum, Iain Boal, Susan Butler,

Yaakov Garb, Beth Horning, Matthew Puma, Rajini Srikanth, Matthew Turner, a discussion group of the Science, Technology, and Values Program, and several anonymous referees have helped me highlight important themes and clarify my prose.

I will not follow the convention of taking responsibility for any "errors"—this places too much emphasis on the author as a revealer of truth. But I will surely have erred in failing to acknowledge some contributors to the development of this project. I invite reminders and rejoinders from you all in the spirit of interrupting patterns of isolation, to which academics are susceptible, especially when finishing "their" books.

I am happy to abide by a different convention, that of ending acknowledgments with a dedication to one's immediate family. My parents, Gilly and John, passed on their disposition to explore more directions than can be held onto securely and to be somewhat unrealistic in trying to discipline that unruliness. My son, Vann, asked me a while back to make the next book one that children would read; I expect him to continue to challenge me to speak to the concerns of people unlike myself. In writing this book and in far more, I have been very fortunate to have Ann Blum as my primary critic and patient guide in the lifelong project of linking the "4H's"—head, heart, hands, and the need for human connections.

Prologue

Overview

Simply put, this book explores complexity and change. To be more specific—although at this point very abstract—I am interested in situations that do not have clearly defined boundaries, coherent internal dynamics, or simply mediated relations with their external context. Such *unruly complexity,* as I call it, arises whenever there is *ongoing change in the structure* of situations that have built up over time from *heterogeneous components* and are *embedded* or *situated* within wider dynamics. In the chapters ahead I explore the significance of unruly complexity in three realms: in ecology and socio-environmental change (in which social and ecological processes are interwoven); in the interactions among researchers and other social agents as they establish what counts as knowledge; and in efforts to feed interpretations of those interactions so as to influence ecological research—or, more broadly, to link knowledge-making, interpretation, and engagement in social change.

I construe social change broadly. Its scope may be as far-reaching as stemming the degradation of some natural resource or redirecting government policies for allocation of funds to different scientific fields (Kaiser 2000a). But social change may also be as local as securing a three-month extension to complete a research project or managing to focus an audience's attention on certain themes. Indeed, the latter is closer to the spirit of this book. I do not provide possible solutions to pressing environmental, scientific, or social problems, nor a comprehensive theory of their causes. My goal—which is ambitious enough—is to stimulate scientists who study ecological complexity and researchers who interpret the ecological-like complexity of scientific change to become more self-conscious and systematic about the ways in which they deal with the unruliness of complex situations. This shift would involve researchers reflecting more *critically*—that is, in relation to alternative possibilities—on their efforts to modify the social and technical conditions in which their research takes shape.

The way in which I promote critical reflection on concepts and practice is to introduce questions and themes that are intended to disturb the conceptual boundaries used by researchers when they focus attention on (supposedly) well-bounded systems and push complicating dynamics or processes out of view. I develop these questions and themes through concrete cases from my own work; these cases open up one to the next in a way that mirrors to some degree the critical reflection I favor. The cases all involve ecological or socio-environmental situations, but their style and content differ according to the intellectual field in which each case is centered—first, theoretical ecology (chapter 1), then philosophy of science (chapter 2), history of science (chapter 3), sociology of science (chapters 4 and 5), socio-environmental studies (chapters 4–6), and eventually, critical reflection on practice (chapters 5 and 6).

The sequence of cases should help researchers and students in this wide range of fields appreciate more acutely the limitations of assuming that ecological, scientific, and social complexity can be delimited into well-bounded systems. My hope is that readers will then take steps—on their own and in collaboration with others—to reconstruct the unruliness of complexity without suppressing it, to link knowledge-making to social change, and to wrestle with the potential and limitations of critical reflection as a means to redirect practice. In the words of Raymond Williams (1980, 83), I want to encourage others not to "mentally draw back [and be] spared the effort of looking, in any active way, at the whole complex of social and natural relationships which is at once our product and our activity."

Historical Origins

Why undertake a project that addresses complexity and change across the three realms of science, interpretation of science, and critical reflection on practice? One answer would be that these realms have always been connected, but concepts and practice are shaped to make them seem separate. This is a position that can emerge only after the book has worked its way through many steps. A shorter answer that might suffice in the meantime derives from the project's historical origins, which can be located at the intersection of two kinds of ecology during the 1970s.

A century earlier, Ernst Haeckel had defined "ecology" as the study of the complex interrelations among animals, plants, and their living and non-living environments (Allee et al. 1949). The meaning of the new term soon stretched to refer to the complex interrelations themselves as well as the scientific study of them. Starting around 1970, "ecology" (and the prefix "eco-") also became associated with actions responding to the degradation of the environment of humans and other species. The array of endeavors that have come under the umbrella of ecology-as-social-action is vast: preventing pollution, ozone holes, global climate change, and other future catastrophes; advocating radical social change, environmental activism, recycling, simpler lifestyles, and unrefined foods; preserving nature, biodiversity, and endangered species; promoting balance and interdependency.

Ecology-the-science promised to help address ecological concerns from a number of angles. Researchers competent in using tools of ecological research could provide technical assistance with particular environmental problems. Systematic environmental analysis and planning might be established so problems could be managed before they became the crises that provoke environmental campaigns. General theories of ecological complexity might enlighten humans about the conditions for more harmonious relations among people and with other organisms sharing our environment.

The rise of ecology-as-social-action, however, also involved a serious critique of the scientific enterprise. The presumption that scientific advances constitute progress was challenged by peace and environmental activists, among others. The destructive effects of science applied, for example, in military technologies and synthetic agrochemicals made it hard to justify the pursuit of knowledge as a good thing for all. The pertinent question was raised: Who benefits from scientific research, and who does not? Such probing exposed science's role in many forms of domination: developed nations over former colonies, military and security branches of the state over dissenting citizens, managers over workers, whites over other races, men over women, and humans over nonhumans. Some people saw science in the service of domination as abuse, not use, of science, but other critical commentators associated it with the nature of scientific inquiry itself. Either way, science was not viewed as unfettered inquiry; instead, specific developments in scientific knowledge began to be interpreted in terms of the social priorities of the governmental bodies, military agencies, corporations, and individuals who sponsored, created, or applied them.

The critique of science also involved positive proposals for alternative processes of inquiry and alternative applications of the products of science. To counter the inherent tendencies of science toward domination—or the recurrent abuses of science in that direction—these alternatives, it was argued, should revolve around cooperation and should not take the contributions of other people or species for granted. Scientists were urged to accept local, democratically formulated input into their research. Even among scientists who insisted on their freedom of inquiry (albeit within parameters set by their funding sources), there was wide recognition of the need to take more responsibility for how the knowledge they made would be applied.[1]

In short, ecology-as-social-action challenged ecological researchers not only to attend to ecological concerns through technical assistance, analysis for planning, or general theories, but also to shape their scientific practices and products self-consciously so as to contribute to transforming the dominant structure of social and environmental relations. In retrospect, I would read in the broad terms of this critique of science an overoptimistic assessment of the potential, on one hand, for the social movements of the 1960s and 1970s to bring about radical restructuring of social relations and, on the other hand, for people to transform their lives accordingly—including, in this context, for scientists to redirect their research. Yet the 1970s critique of science was a key aspect of the context in which I first began to engage with the complexities of environmental, scientific, and social change together, as part of one project. The challenge I take up in writing this book, then, is to build on the historical and personal origins of the project and to convey its subsequent evolution in terms that help other researchers engage with such complexities in the context of the early twenty-first century.

Conceptual Exploration: An Autobiographical Narrative

My decision to study ecology during the early 1970s stemmed from environmental activism in Australia that ranged from a collaboration with trade unionists opposing the construction of an inner-city power station to street theater exposing fraudulent industry-sponsored recycling plans (Whole Earth Group 1974). Ecology-the-science was the recommended choice for college students who sought programs of study in which to pursue their interests in ecology-as-social-action—if indeed any other choices were available. I hoped my studies would lead to some kind of career that would take me beyond responding to one environmental issue after another and instead allow me to contribute to planning that prevented future problems from emerging. I also hoped that understanding how to explain the complexities of interactions in life would lend support to less hierarchical and exploitative relationships, both within society and among humans and other species.

I had brought a mathematical disposition to my studies in ecology, so I undertook projects that advanced my skills in quantitative analysis and mathematical modeling. I was excited to learn that some biologists and mathematicians were creating a specialty called theoretical biology (Waddington 1969). This discovery was still fresh when I took a course for which E. C. Pielou's (1969) text on mathematical ecology was assigned. In the introduction, she noted that organisms come from a range of species; that within any species, they differ in age, sex, genetics, experience, and so on; and that any particular individual changes over its lifetime. Any situation an ecologist might study is continually altered by births and deaths, by migratory exchanges with other places, and by seasons and climatic change. Even so, ecological regularities persist long enough for most people to recognize some order, such as an oak-maple forest or the sequence of plants encountered as one moves inland from the seashore (Pielou 1969, 1). The processes involved could be simply described, yet the combination of them seemed theoretically challenging—how could ecologists account for order arising out of such complexity? (to be continued ...)

◆ ◆ ◆

Framework, Audiences, and Positioning

My exploration of the question about theorizing order in ecological complexity has opened out into inquiries in socio-environmental studies, interpretation of science, and critical reflection on practice. I have examined the ways in which researchers in diverse intellectual fields address—or discount—the heterogeneity, embeddedness, and ongoing restructuring that make complexity unruly. I have questioned the assumption that environmental, scientific, and social complexity can be partitioned into well-bounded systems that can be understood or managed from an outside vantage point. Instead, I have looked for concepts and practices that would help researchers treat boundaries of many kinds as problematic—including the boundaries between science, interpretation, and engagement.

The chapters ahead represent steps in the development of a framework, made explicit in the final chapter, that integrates conceptual, contextual, and reflexive angles on the practice of researchers. In brief, the framework

- exposes the hidden complexity of the simplifications that various fields use to focus attention on supposedly well-bounded systems;

- takes an expanded view of ecological and socio-environmental research in which inquiry is embedded in interactions among researchers and other social agents as they establish what counts as knowledge;
- extends that perspective on the embeddedness of knowledge-making so that it also applies to research that interprets science (in the fields, for example, of philosophy, history, or sociology);
- locates interactions to establish knowledge within a larger realm where researchers pursue social change through diverse and often modest practical choices;
- invites researchers to address self-consciously the complexities of the situations they study *and* the social situations that enable them to do their own work; and
- makes space for *conceptual exploration*; that is, for playing with themes and models so as to open up questions in broad terms that might transfer across different fields while, at the same time, keeping the limitations of such themes and models in view.

I hope that an exploration of "unruly complexity" resonates with some of the concerns of a diverse range of readers: ecologists and socio-environmental researchers, modelers and theorists of complexity in biological and other systems, interpreters of science, and educators and activists in environmental or scientific politics. Yet the audience I envisage is defined not by field or discipline as much as by three qualities: an interest in exploring new propositions, themes, questions, or framings and seeing how these might be adapted to their own inquiry; a sense that disciplinary boundaries (for example, between science and interpretation of science) give them trouble in their work; and a disposition to reflect on the conceptual and practical choices they have made in relation to alternative possibilities, past and future.

To draw readers into issues outside their field or from an episode or period in the past and to keep them moving through unfamiliar terms and disciplinary styles, the book includes an autobiographical narrative and several other expository aids: puzzles to open the cases, a glossary, a summary of themes and questions, and thematic endnotes.

The *autobiographical narrative,* which began above, bridges the chapters and highlights the conceptual connections between cases from different fields. It also serves a number of other functions. The narrative allows me to downplay—as most storytellers do—side branches and intersecting story lines, so that readers can see the logic of the book's overall progression even when the cases take them beyond their specialties. Readers are welcome to read this story in its entirety first for a preview of how the diverse components of my project come together. Similarly, any reader who balks at the technical detail of, say, the early chapters on ecological modeling may use the narrative to move ahead to chapters with which they are more comfortable and return later when they start to see the connections across the different fields. (For additional help digesting the concepts and connections, refer to the *glossary* and *summary of themes and questions* included in the end matter.)

The narrative also serves as a reminder that the episodes in the text are drawn from *certain periods of time* in the work of *one* person. This caveat means that readers should not expect a direct payoff in relation to the current problems in any of the fields addressed—I would not have arrived at a framework spanning the three realms of complexity if I had followed the pathways that branched off after each episode or had tried to translate my cases into the present terms and inquiries of each field. (The extended *thematic endnotes,* which acknowledge subsequent developments and related projects, should help readers relate my approach to those of other authors.) Notwithstanding this caveat, the cases from the past are not meant to be read as dated episodes; each exposes issues that, I believe, still call for attention because they were not well resolved or recognized when researchers in that field moved on. (The *puzzles* are included at the start of each case in order to draw readers into—or back into—those issues.)

The most important function of the narrative, however, is to model the journeying, opening up of questions, and critical reflection on practice that I aim to stimulate in others (Taylor 2002b). Readers could well ask me for evidence or argument that conceptual exploration and critical reflection produces better or more relevant answers for their field. I ask them instead to view this narrative, which highlights the partiality, particularity, and developmental character of my contribution, as an invitation to reflect on their own paths and positioning over time. In this spirit, readers might bounce off the themes I develop in the book and articulate the contrasting or similar themes they have applied themselves when addressing the complexities of the situations they study and the conditions that enable them to do their work.

About my own positioning: I remarked earlier that ecology-the-science could address ecological concerns at three levels: through technical assistance with particular environmental problems, systematic analysis and planning, and general theories of ecological complexity. Interpretation of science can also be arrayed along an equivalent spectrum. Clearly this book lies at the general end, with an emphasis on questions and themes that

I believe can help researchers address complexity in ecology, interpretation of science, and critical reflection on practice. Systematic analysis and planning are also considered, but mostly in ways that are critical of technocratic ambitions and that promote reframing of analysis and planning along participatory and reflexive lines.[2] The worthy demands for assistance with particular environmental and scientific problems are not the subject of this book.

I believe that conceptual exploration is valuable for researchers trying to deal with environmental, scientific, and social complexity. But let me acknowledge at the outset that contrasting approaches, especially learning from long-term engagement in particular controversies or in larger social mobilizations,[3] keep me questioning my emphasis on such abstract concepts as heterogeneity, embeddedness, and ongoing restructuring. I see conceptual exploration as the area in which I have been best able to make a contribution, but I look forward to dialogue that keeps such tensions active and productive—that stretches what I am able to offer in the text of a book.[4] Indeed, the framework I develop in the chapters ahead makes a special place for conceptual moves that open up issues about addressing complexity, but do so in ways that point to further work that needs to be undertaken to deal with particular cases.

▲

My undergraduate studies had raised the theoretical question of how ecologists could account for the order arising out of complexity, but the jobs I applied for after graduation were more practically oriented. Environmental planning scarcely existed in Australia in the mid-1970s, and I found employment in agricultural research. My first job was to extract patterns from data about the complexity of interactions between plant varieties and field conditions in large crop trials. My second job involved modeling the economic future of an irrigation region suffering from soil salinization (a project analyzed in chapter 4). To my frustration, the government sponsors of the salinization study turned out to be interested in only a small subset of the factors and policies potentially relevant to the region's future. This experience in analysis and planning led me to seek opportunities for self-directed inquiry in ecological theory. At the same time, the experience motivated me to explore ways in which social influences could shape ecology and environmental science in less constraining ways.

My interest in understanding science in its social context had already been stimulated by the advisor of my undergraduate thesis in ecological modeling, Alan Roberts, a physicist who also wrote about environmental politics and the need for the self-management of society (Roberts 1979). From Roberts and others, I was learning that through the course of history, all kinds of social lessons had been read from nature (Williams 1980). It would be better to argue directly for, say, cooperative, decentralized social relations than to put forward some account of ecological *complexity to justify them. Nevertheless, I could still envisage research on complexity challenging the simple scientific themes that were often invoked in support of social inequalities and exploitation of nature (Science for the People 1977). As I was finishing the salinization study in 1979, I learned that two biologists in the United States whose theoretical work I already knew and valued, Richard Levins and Richard Lewontin, saw their scientific work as a political project (Levins and Lewontin 1985; Taylor 1986). I sought an opportunity to study with them. This would draw me away from environmental activism in Australia, but this leave—which has extended longer than I could have imagined—would provide space to focus on questions around conceptualizing life's complex ecological context and to begin to take up questions of conceptualizing science's complex social context . . .*

▲

Part I

Modeling Ecological Complexity

Ecological complexity poses challenges to conventional scientific ways of knowing. Ecology is not like thermodynamics, in which complexity can be simplified through statistical averaging of large numbers of identically behaving components. Moreover, whereas progress in the physical sciences depends greatly on controlled experiments in which systems are isolated from their context, this strategy is not so clearly appropriate for understanding organisms in a context of interactions with a multiplicity of hazards and resources distributed in various ways across space and time. At the same time, analysis of observations from nonexperimental situations is beset by circularity—ecologists need to know a lot in advance about causal factors before they can design methods of multivariate data analysis capable of revealing the effects of those factors (see chapter 2, note 10).

If ecological complexity does not lend itself well to statistical simplification, experimental control, and multivariate data analysis, it is fair to ask whether any general theories of its structure could apply. During the 1960s and 1970s many academic ecologists, especially in the United States, had thought that such theories were indeed possible. In systems ecology, complexity was analyzed in terms of the nutrient, energy, and information flows among the living and nonliving entities that make up the entire ecosystem (to be discussed in chapter 3). In community ecology, analyses focused on a part of the ecosystem; namely, on some group of interacting populations or ecological community.[5] Theoretical propositions concerned population sizes and distributions and their regulation through interspecific interactions—chiefly competition for limiting resources. Elegant, widely applicable principles of ecological organization were sought. Robert MacArthur, following the lead of his teacher G. Evelyn Hutchinson, was a leading proponent of expressing such principles as verbal or mathematical models: "Will the explanation of these facts degenerate into a tedious set of case histories, or is there some common pattern running through them all?" (MacArthur 1972, 169). By using mathematical equations to focus attention on certain entities and relationships in ecology, MacArthur, Hutchinson, and other ecologists encouraged mathematicians, including a number of my teachers in Australia, to try their hand at ecological theorizing.

In the early 1980s—the time when I began doctoral studies in the United States—ecologists of a particularistic bent were vigorously questioning ecological principles and expressing skepticism about the possibility of general ecological theory. Daniel Simberloff (1982), for example, argued as follows: Many factors operate in nature, and in any particular case, at least some of them will be significant. A model cannot capture all the relevant factors and still have general application. Instead, Simberloff contended, ecologists should intensively investigate the natural history of particular situations and test specific hypotheses about those situations experimentally. Ecologists may be guided by knowledge about similar cases, and they may end up adding to that knowledge, but they should not expect their results to be extrapolated readily to many other situations.

The tension between the MacArthurian tradition and the newer critiques stimulated me to clarify the relations intended between models and the reality to which they refer—the subject of chapter 2. Yet, despite the particularistic challenge to models, I remained interested in fundamental questions in ecological theory. One such question was how explanations that involve interacting causes can be extracted from ecological patterns and data (see chapter 2, note 10). Another question—which underlies the two parts of chapter 1—concerned the consequences of defining boundaries between the outside and inside of a system when ecologists attempt to account for ecological structure or organization . . .

1

Problems of Boundedness in Modeling

Ecological Systems

A. The Construction of Complexity

What have ecological theorists learned about the relationship between complexity and stability?

MacArthur (1955) proposed that "the amount of choice which the energy has in following the paths up through the food web is a measure of the stability of the community" (fig. 1.1). This formulation gave a mathematical touch to an old intuition that the diversity or complexity of interrelations among organisms ensures harmony or stability in the order of nature (Egerton 1973). By 1975, however, Goodman, after reviewing the voluminous literature and sorting through diverse definitions of complexity and stability, had concluded that the few well-defined empirical and theoretical studies tended to contradict the expectation that complexity promotes stability (Goodman 1975; see also INTECOL 1974). Ecologists continued to tease apart the different meanings of the terms and examine a variety of complexity-stability relationships (Pimm 1984, 1991), but by the late 1980s, most had become skeptical about any robust relationship ever emerging. For example, in a poll of the members of the British Ecological Society for its 75th anniversary symposia, "the diversity-stability hypothesis" ranked only 35th in a list of the 50 "most significant" ecological concepts. These

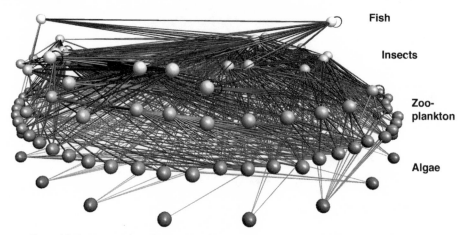

Fish

Insects

**Zoo-
plankton**

Algae

Figure 1.1 A representation of one aspect of ecological complexity. Trophic or feeding relationships in Little Rock Lake, Wisconsin, are indicated by a link between nodes that contain all species (or other taxonomic groupings) that share the same predators (links above the node) and prey (links below the node). (Figure provided by N. Martinez; see Martinez 1991.)

ecologists deemed "ecosystem fragility" (ranked 10th) a much more "significant" concept (Cherrett 1989).

What exactly could ecologists claim to have learned in turning away from the complexity-stability issue by the end of the 1980s?[6] Was the lesson that change, rather than constancy and stability, is the normal state of ecology (Botkin 1990)? Hardly—change, process, and dynamics were central concepts for the founders of the discipline (Kingsland 1995, 9–24). Even Frederic Clements (1874–1945), whose name is associated with the idea of stable, self-sustaining climax communities of plants, was primarily concerned with explaining ecological succession—the changes at some site over time that led predictably, he thought, to some such climax (Clements 1916). And, in the writings of Clements's leading critic, Henry Gleason (1882–1975), nothing was static. Ecologists, Gleason proposed, should analyze the shifting associations of individual species as they respond to variable environmental conditions and view succession as "an extraordinarily mobile phenomenon" (Gleason 1927).

No, the ubiquity of change was not a new insight for ecology. Admittedly, given that environmentalists still invoke the age-old idea of a balance of nature, there was room still for scientific debunking of this concept (Botkin 1990). The complexity-stability question, however, need not be disposed of in a package with the balance-of-nature myth. Nor need it be reduced to questions of productivity and constancy of numbers in mixtures of plants (Tilman 1999). Instead, I believe, attempts to find a stability-complexity connection can be viewed in relation to the long-standing problem of accounting for structure and organization as well as the possibility of change—in short, for ongoing restructuring. In this light, let me reconsider the search during the 1970s and 1980s for a significant relationship between complexity and stability, in which the definition of both terms varies among authors. The lesson will not be that ecological interactions are without any structure or constraints on change. A myth of "anything goes"—that all disruptions are qualitatively alike—would offer no better guidance for human intervention in (or within) nature than the myth of harmony and balance.

Stability of Simple Multipopulation Models

Let me conduct this review in the spirit of MacArthur; namely, by exploring the qualitative behavior of simple mathematical models in the hope of finding some theoretical unity among the disparate facts of ecology. The terms of the models will be the numbers of organisms in each population and the interactions—competition, predation, parasitism, and so on—that influence the growth or decline of those numbers. Contingent circumstances that could obscure any regularities will not be considered (Kingsland 1995, 176–205). These simple models, which have few explicit biological assumptions, are intended—using the analogy of MacArthur's physicist-turned-theoretical-ecologist successor, Robert May—to be like the perfect crystals of solid-state theory in physics. They are gross simplifications and obtain nowhere, but investigation of the deviations of reality from the ideal will suggest new features to add to the initial models—or, as often will be the case, to new formulations of the question at hand. The justification for this approach is that if we were to write down at the outset a model with many specific biological features, it would be difficult to establish whether its behavior depended on the specific features or simply resulted from the basic dynamics common to many other models (May 1973, 10–12). (In chapter 2 this "light-on-data" strategy of modeling is compared with other strategies of modeling.)

At first sight, the impression that nature is complex and nature is stable admits of two broad interpretations:

1. Nature is stable because *complexity in general* promotes stability.
   This interpretation is consistent, for example, with the observation that

monocultures are much more vulnerable to pest outbreaks than are
diverse, natural floras and faunas.

2.   Nature is stable because of its *particular complexity*. This interpretation
is consistent with the idea that human disturbances can threaten nature's
finely tuned checks and balances.

Let me start with the first view. Could stability as a consequence of complexity be a general property of complex systems? Beginning around 1970,
this possibility was examined through computer-aided investigations of
mathematical systems governing the rate of change of many interacting
components (Gardner and Ashby 1970). For the purposes of ecological theory, the systems could be thought of as ecological communities and their
components as populations of interacting species (May 1972). May's
approach was to compare many samples, each made up of many model
communities. Within any sample, each community would have the same
complexity, in the sense of the number of populations, strength of interactions among populations, and interconnectedness (the proportion of pairs
of populations actually interacting). The communities would vary only in
the arrangement of positive, negative, and zero interactions. The equations
relating rates of change of populations to their interactions were bare of
biological detail (such as spatial location or individual variability) and were
not constrained to fit any recognizable "trophic" structure (such as food
chains from plants to carnivores). The equations were simple enough to
calculate the equilibrium or steady state population sizes of each community and to determine the community's stability in the sense of tendency to
return, when perturbed slightly, to that equilibrium.

    With the help of their computers, modelers such as May discovered that
stable communities become more rare in the samples with higher complexity. Moreover, for the stable communities to be of biological relevance, they
must also be "feasible"—that is, all populations must be positive—and the
chance of feasibility drops rapidly with the size of the community (Roberts
1974). In short, complexity almost ensured instability (fig. 1.2).

    Responses to the "complexity-instability" results were varied, and each
response reformulated the issue in some way: In what sense and to what
extent are real ecological communities stable? Do real communities keep
below the limiting levels of complexity and strength of interactions suggested by the models? Might complexity promote stability in models whose
equations captured more biological detail? If complexity in general does
not promote stability, what are the "devious strategies which make for stability in enduring natural systems" (May 1973, 174)?

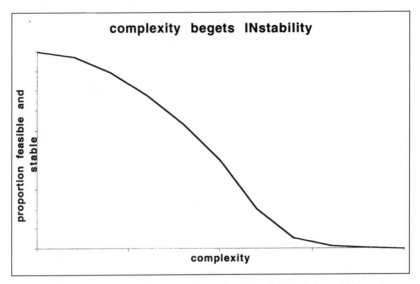

Figure 1.2 The proportion of model systems that are feasible and stable declines with increasing complexity (i.e., number of populations, strength of interactions, and degree of interconnectedness).

Goodman's (1975) review concluded that experiments and observations complement the results from modeling. Complexity, when measured by various indices of diversity (Pielou 1975), bore no consistent, let alone positive, relationship to stability, as assessed by the degree of fluctuations of populations or simply by their persistence. Subsequently McNaughton (1977) reported a positive relationship for grasslands. But it was ecological modeling, more than new empirical evidence, that kept the issue alive into the 1980s. Or, one might say, it was the different formulations of ecological modelers that gave the complexity-stability issue new lives.

Saunders and Bazin (1975), for example, realized that if there is any biological basis to the idea that complexity and stability are related, it must have been derived from actual ecological communities. For these communities to have been observed, they could not have been ephemeral. Saunders and Bazin, therefore, considered only model communities having a stable equilibrium and asked whether such communities were more resistant on average to perturbations when more complex (fig. 1.3). (Their question, it should be noted, falls under the second broad interpretation; namely, that nature is stable because of its particular complexity.) They discovered, however, that the speed of recovery from perturbation decreased with complexity. Pimm (1979) extended this line of inquiry by investigating stability after a much larger perturbation; namely, deletion of one population

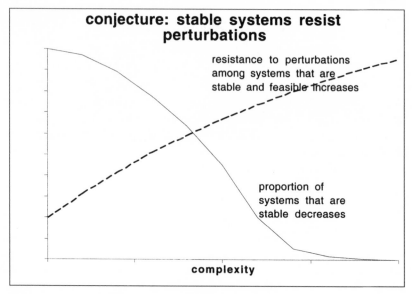

Figure 1.3 The possibility that model communities that have a stable equilibrium are more resistant on average to perturbations when more complex.

from a stable model community. He discovered that the greater the inter-connectedness of populations, the less likely the community was to be stable after the deletion of a population. In short, the reformulation from the proportion of stable systems to the degree of resistance of stable systems to perturbation did not seem to illuminate why complex ecological communities would be stable.

Suppose that the stable complex communities examined by modelers had turned out to be more stable, in some sense, than stable simple communities. Theory would still have had to explain how, given the rarity of such complex communities in the samples of May and others, any of them could come to exist. In the spirit of May's call to expose the "devious strategies" promoting stability, various modelers sought to elucidate the special conditions or structures that ensure that complexity promotes stability, or, at least, ensure that complexity does not make stability very unlikely.

McMurtrie (1975), for example, showed that complex communities are more likely to be stable if nonzero interactions between populations form long cycles from one population through others, then back to the first population. Siljak (1975), following the lead of Simon (1969), suggested that stable communities can be "decomposed" or partitioned into blocks having strong intra-block interactions and weak inter-block interactions. This would mean that in stable communities, intraspecific interactions over-shadow interspecific interactions. Features such as low efficiency of assimilation of consumed food, absence of loops of the kind "X eats Y eats Z eats X," and low predation efficiency at low densities were also shown to enhance the complexity-stability relationship (DeAngelis 1975; Lawlor 1978; Nunney 1980; but see Abrams and Taber 1982). Other authors examined the consequences of particular features, such as time lags—between a population change and its effect on an interacting population—and fluctuating environments (May 1973), more complex forms of the equations (Pomerantz, Thomas, and Gilpin 1980), and inclusion of nutrient uptake and cycling (DeAngelis 1992).

Cycles, partitioning, and so on may enhance the possibility of stability in model communities. Yet for each addition that ecologists make to the list of stability-enhancing mechanisms, nature becomes less obliged to employ any particular one of them. Furthermore, even if ecological data display some of the features found in models (Pimm 1984, 1991), it does not follow that they exist in nature *because* of their stabilizing effect. A stronger case for any mechanism is made if the modeler can account for how nature arrives at the special structures, or how populations violating the special conditions are eliminated. That is, modelers need the dynamic evolution of their model communities to produce the feature in question. This means that modelers have to resist pressure for their models to look realistic to other ecologists at the outset. When modelers impose some biologically realistic constraint on their model communities in advance, they lose the chance of explaining it dynamically.

If dynamic evolution is required, then the question of how any stable complex community exists is modified: In what ways does nature arrive at the arrangements that persist longer than those that preceded them? Whereas May's complexity-promotes-instability result assumes that complex ecological communities are outcomes of a sampling process, most actual communities are the result of some process of development over time. Many authors have suggested that natural selection might operate in a way that modifies interactions among populations in a direction that enhances the stability of communities (Lawlor and Maynard-Smith 1976; Roughgarden 1977; Saunders 1978; Lawlor 1980). However, early theoretical investigations of a very simple community (one predator–one prey) suggested that this is not necessarily true (Rosenzweig 1973), and subsequent modeling of more complex communities of competitors indicated that interactions become progressively weaker until only a few interacting populations dominate (Ginzburg, Akcakaya, and Kim 1988).

This continuing series of negative or not-so-compelling complexity-stability relationships might lead one to conclude that there is, as Goodman (1975) intimated, nothing there to be discovered. This is the position adopted by DeAngelis and Waterhouse in their important review (1987). To these authors, May's complexity-instability result indicates that a "potential biotic feedback instability [is] inherent in complex natural systems" (DeAngelis and Waterhouse 1987, 5). (By the 1980s, this result had become conventional wisdom, a status that persists, even among the plant ecologists who have revived interest in diversity: see note 6; but see May 2000.) Noting the limited success of different attempts to counteract this instability, DeAngelis and Waterhouse concluded that one should not "base any theory of ecological communities solely on the notion of inherent ecological stabilizing mechanisms" (1987, 9).

DeAngelis and Waterhouse's review of ecological modeling mapped even more paths than I have so far (fig. 1.4). They started, as I did here, from an "equilibrium" view, in which communities move toward or away from a steady state. Attention to disruption from internal feedbacks led ecologists to emphasize "biotic instability" (path a in the figure). Similarly, environmental disturbances that are irregular (or "stochastic") led to "stochastic domination" (path b). To account for the persistence of communities despite these disruptions, several paths have been taken: exploration of potential stabilizing mechanisms (as I have discussed; path 1), studies of disturbance patterns that might interrupt any process that eliminates an interacting population (path 2), and studies of biological mechanisms that might compensate when populations have been driven to low numbers (path 3). But to DeAngelis and Waterhouse, the most promising approach entails a "landscape" view, in which a community may persist in a landscape of interconnected patches even though the community is, because of instability and environmental disruption, transient in each of the patches (paths 4 and 5). Although I endorse most of DeAngelis and Waterhouse's interpretations, I want to draw attention to another path that leads to the landscape view, which they, and most other modelers, have overlooked or discounted (path c and 6). To introduce this path, I need to return to the question of how nature arrives at the arrangements that persist.

A Constructionist View

Real ecological communities develop over time through succession—that is, the addition, growth, decline, and elimination of populations. Several modeling studies have shown that although stable, feasible model commu-

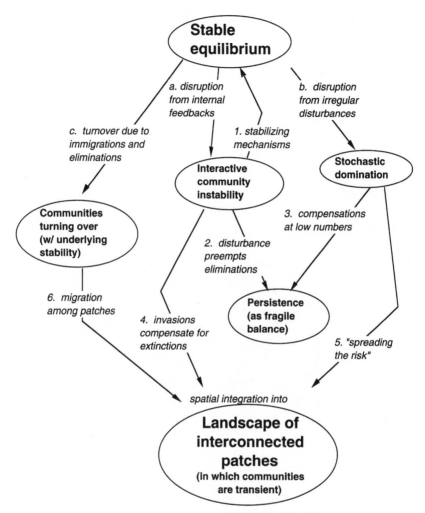

Figure 1.4 Developments in theorizing ecological complexity and stability. Paths with letters are sources of instability. Paths with numbers are potential stabilizing mechanisms. (Adapted from DeAngelis and Waterhouse 1987; see text for explanation of additional paths c and 6.)

nities may be extremely rare as a fraction of the communities sampled, they can be readily *constructed* over time by the addition of diverse populations from a pool of populations and by the elimination of populations from communities not at a steady state (fig. 1.5). (Tregonning and Roberts [1979] made important initial contributions, which I followed; see Taylor 1988b and references therein.) I call this insight about complexity-stability theory a *constructionist* view.[7]

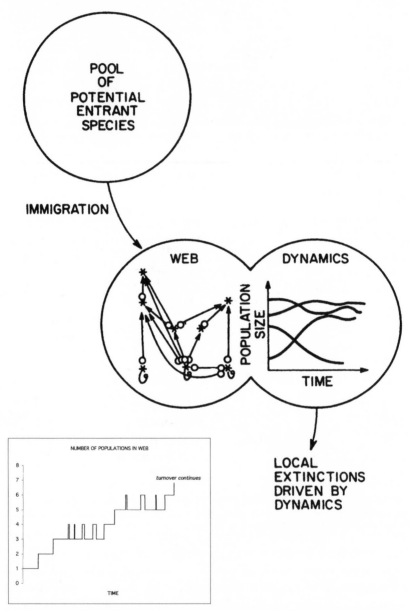

Figure 1.5 The construction of stable, feasible complex systems through addition from a pool of potential entrants and elimination due to the dynamics of the resulting community. The number of species in the web at any time (bottom left graph) reflects additions and eliminations. (Adapted from Taylor and Post 1985.)

Construction is reminiscent of ecological succession, which field ecologists have never forgotten (Gray, Crawley, and Edwards 1987), but its implications have not been well appreciated by the theoretical ecologists mentioned in the preceding review of the stability of simple multipopulation models (though see Post and Pimm 1983; Nee 1990). Similarly, the constructionist view has not been explored in research on "assembly" rules, which are derived from empirical and statistical analyses of species co-occurrence (Diamond 1975; Weiher and Keddy 1999; though see Drake et al. 1999). And it is missing from the debates in plant ecology over the effects of diversity on productivity (Taylor 2000a). So let me tease out some of the ways in which construction has significance for ecological modeling and theorizing (Taylor 1989a):

1. **Ecological complexity may persist through ongoing community "construction and turnover."** If a community that is stable (in the sense of resisting perturbations from equilibrium) has been constructed over time by addition and elimination of diverse populations, it is also a community that has been constructed from and supplanted a previous stable constructed community. It may, in turn, be supplanted by another community. Given this turnover, ecological complexity may persist at the expense of the transience of any community. This conclusion is an ironic twist on the search for a complexity-stability relationship.

This conclusion may sound similar to that of DeAngelis and Waterhouse, but it was derived by different reasoning. Admittedly, since a primary source of potential immigrants or invaders would be neighboring patches, the constructionist view can be seen as a variant of their landscape view. (Other sources might be refuges in which species persist undetected at low abundances and the seed bank.) Remember, however, that the transience of communities in a patch is caused by invasions, not by intrinsic biotic instability. In the constructionist view, stability underlies any transient configuration of populations. Roberts and Tregonning's (1981) analysis of their constructed communities found that many nested or overlapping stable subsystems make up the stable community as a whole. In short, communities may be susceptible to new invasions and restructuring at the same time as they are stable with respect to perturbations from equilibrium or random eliminations of populations.

Construction and turnover might seem to mean that the buildup of complexity could go on without limit—that as long as communities are subject to invasions, populations would tend to be added. This growth is, however, checked by another factor. Constructed communities are stable with respect to perturbations and population eliminations, but relative to their simpler precursors, the more complex ones tend to be less stable. As shown in figure 1.6, communities would then be expected to approach some balance between addition by invasion and elimination through lowered stability (Robinson and Valentine 1979).

2. **The range of mathematical possibilities for models can be extended.** Under the constructionist view, investigations of complexity in ecological models should go beyond the conventional analysis of the current configuration or "morphology" of that complexity and include the historical development of that complexity as it is embedded in its wider spatial context. The morphological approach takes into account the rarity of stable communities in the space of possible arrangements of interactions and the mathematical intractability of complex models, and this means that it has to focus on analysis of communities with only a few variables, or with many identical variables that yield a statistical simplicity. It considers a given type of model to be an unlikely candidate for representing natural communities if the model turns out to be stable only in a small region of the space of possible arrangements. In contrast, under a constructionist approach, it is possible to arrive at models located in this small region,

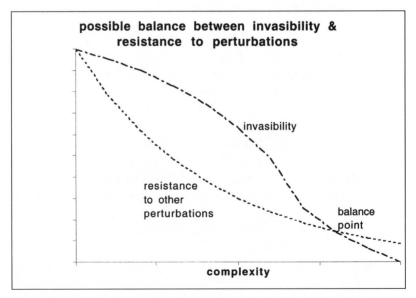

Figure 1.6 How the construction and turnover of communities might approach some balance between additions by invasion and eliminations through lowered resistance to other perturbations.

which means that likely models cannot be rejected on the basis that the region of parameter space they occupy is too small. Moreover, the model's dynamic evolution is tracked by computer simulation and is not determined by direct mathematical analysis, so it is possible to consider model communities that have more biological detail. This frees the modeler's imagination to explore arrangements whose analysis is less mathematically tractable, such as models that consider nutrient and energy flows in addition to population growth and declines (Taylor and Post 1985).

3. **Persistence of complexity does not necessarily require devious strategies.** The constructionist approach provides new perspectives on some of the stability-enhancing results mentioned earlier. Contrary to Siljak's emphasis on complex communities being partitionable (decomposable) into blocks, richly interacting communities can be readily produced by development over time (Taylor 1988a, 1988b). Although this result does not rule out the possibility that ecologists will find that natural complexity is partitionable, theoreticians should take note that nature—or any complex entity, such as a society, for that matter—does not have to restrict itself to hierarchies of weakly interlinked components (Simon 1969; Johnson 2000). Similarly, model communities lacking McMurtrie's loops can be readily constructed; so can communities well above the critical levels of complexity defined by May (Taylor 1988b). Indeed, although "devious strategies" certainly abound in ecology, the persistence of ecological complexity might be largely explicable without recourse to any of them. Ongoing turnover of populations may be all that is needed.

4. **Complexity can be constructed in ecological time without natural selection.** Construction can occur in ecological time—that is, without requiring genetic fine-tuning of population interactions. Drake discovered that altering the sequence of colonizing populations may result in significant differences in the communities constructed (Drake 1990), an effect he replicated in micro-ecosystem experiments (Drake 1991; see also the experiments of Robinson and his colleagues: Robinson and Dickerson 1984, 1987; Robinson and Edgemon 1988). These models and experiments show that historical ecology need not be evolutionary—at least, not in the sense of requiring genetic change and adjustment by natural selection. Ecological and coevolutionary processes and timescales can be separated (Herrera 1986).

5. **Complexity constructed in ecological time depends on its spatial context.** Because the pool of populations entering the modeled community is drawn primarily from surrounding communities, the relationships among

places or patches need to be included (Drake 1990). As DeAngelis and Waterhouse's landscape view emphasizes, historical ecology is also necessarily ecology in a spatial context.

6. Complexity is better conceived in terms of intersecting processes than well-bounded systems. The constructionist view, like the landscape view, suggests that persistence at one scale in time, space, or degree of aggregation can result from the intersection of processes at other scales. Such processes include the addition of populations from surrounding patches and local elimination. Suppose one's attention moves down to the scale of the constituent processes. With modest changes in conditions, the very same processes may yield a range of outcomes other than persistence—perhaps a mosaic of transient communities continually shifting over time and space. Persistence or steady state would then become just one of a range of ecological situations to be explained.

The idea of shifting or indeterminate boundaries, in turn, invites ecological theorists to question not only the equilibrium view, but also their reliance on a strong concept of system. Of course, the term *system* may be used simply to denote that there are many elements interacting, but systems as they are modeled by mathematical equations are usually entities with well-defined boundaries and coherent internal dynamics governing their development (see the central circle in fig. 1.5). Communities in the models I have discussed above are systems in the latter, strong sense. To the extent that communities exist and persist, the property of system-ness should, under the constructionist view, be a contingent outcome to be explained, not a starting point for theory and modeling (Taylor 1992a).

◆ ◆ ◆

Ecological theory has traveled a long way along the various paths leading from MacArthur's proposition that a choice of energy pathways confers stability. The constructionist view indicates that the persistence of complexity, while not ensured by complexity itself, is not as puzzling as the result that complexity begets instability. But the modeling work I have reviewed also reintroduces historical contingency and spatial context, which, ironically, helps undermine the initial MacArthurian aspirations for general theory about ecological patterns. In fact, the situation is even worse. If complexity is reconceived as a series of temporary outcomes of intersecting processes operating at a diversity of scales, then theorists have barely begun to account for ecological complexity.[8]

Or should I say that the situation is even *better*? Important perspectives have emerged from exploring how to theorize the simple observation that complexity persists. A middle ground has been opened up between the order of Clements's climax communities and the flux of Gleason's shifting associations. A challenge has been brought into sharper focus: namely, to address complexity that has not only structure but also a history of structuring and ongoing restructuring.

B. The Hidden Complexity of Simple Models

A series of attempts to understand the structure and stability of entire ecological systems followed MacArthur's 1955 proposition (with which section A began). An alternative route to understanding ecological complexity, however, would be to derive basic principles about the interactions of a subset of the ecological community—principles, say, about the separation of resource requirements of similar competing populations ("niche partitioning"). Then, if these principles were combined with other principles, an explanation of the entire complexity might eventually be built up. MacArthur, in fact, championed this approach more than he did the direct assault on the problem of complexity.

Particularist approaches called into question the possibility of any principles that could be usefully generalized across ecological situations (see the introduction to this chapter). A different source of doubt, which gained momentum by the end of the 1980s, derived from some theorists attending to "indirect interactions"—effects mediated through the populations not immediately in focus or, more generally, through "hidden variables" (see reviews by Strauss [1991] and Wootton [1994]). Indirect interactions might confound principles derived from observations of direct interactions among populations. To explore this concern, let me begin by presenting a puzzling anomaly. The resolution of this anomaly does indeed disturb conventional interpretations of models that refer to only a few populations when those populations are embedded in naturally variable and complex ecological situations.

Why were half of the interactions in a community of competing protozoans predator-prey relations?

Field ecologists are perennially skeptical about textbook mathematical models, but Vandermeer (1969) fitted his laboratory data on four competing protozoan populations to relatively simple equations and used them to make correct predictions about which populations could coexist. I noticed, however, an anomaly in Vandermeer's results that he and subsequent discussants had not addressed: three of the six pairs of interactions between the competitors had values that were positive-negative (fig. 1.7). One would expect positive-negative values for predator-prey relations, not for competitive interactions. Were these interactions actually predator-prey? Indeed, were those pairs with negative-negative interactions actually competitors? How can the values Vandermeer derived be understood and related to the actual ecological relationships among his protozoan populations?

An obvious response might be that Vandermeer's model was inappropriate or inadequate, so let me examine that possibility first. The interpopulation interaction values he derived for his four protozoan species came from fitting the observed changes in population sizes or trajectories to a model of the following form:

Model 1: Generalized Lotka-Volterra (GLV) model
　　　　 Per capita rate of change of population X =
　　　　 intrinsic growth rate for X + self-interaction within the Xs +
　　　　 sum of the effects of the other populations on X

where the first term is a constant, the second is a constant times the size of population X, and the interpopulation terms are constants times the sizes of the other populations.

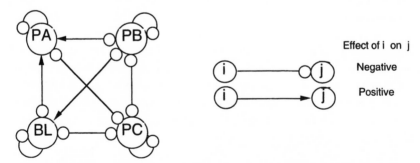

Figure 1.7 Community interactions reported by Vandermeer (1969). PA = *Paramecium aurelia*, PB = *Paramecium bursaria*, PC = *Paramecium caudatum*, BL = *Blepharisma* sp. Three of the six pairs of interactions between the competitors are positive-negative. The symbols for positive and negative effects are borrowed from loop analysis (Puccia and Levins 1985), but otherwise this case has no connection to that approach.

Vandermeer estimated the intrinsic growth term and the self-interaction term from isolated population growth experiments and the interpopulation interaction terms from two-population experiments. Contrary to the widely held opinion that the GLV is a poor ecological model, the fit for Vandermeer's four-population microcosms was fairly good and gave qualitatively correct predictions about coexistence of populations (Vandermeer 1981).

Given that Vandermeer's model fits his observations well, how can we explain the anomalous (– +) interaction values between the competing protozoans? First, note that Vandermeer's equations did not specify all the components of the community. Each day during his experiment he removed a sample from his experimental tubes and added an equal volume of culture medium containing bacteria. The bacterial populations were alive and able to grow until consumed by the protozoans. They had dynamics of their own not referred to in the equation above. In fact, it is possible that the protozoan populations were affecting each other only through their shared bacterial prey. If all the fitted interactions had indicated competition, the unspecified components might not have been cause for concern—the protozoan populations could be described as exploitative competitors. Yet the interactions were not all competitive.

Second, notice that the observed behavior of the protozoan subcommunity—the full community minus the bacteria—was fitted with a model containing interactions only within that subcommunity. Because there was no direct reference to the relationships with the hidden part of the community, the fitted interaction values had to incorporate these other indirect relationships, if they existed. Let me call the fitted interactions *apparent interactions* and use this term whenever ecologists attempt to specify the ecological dynamics of a subcommunity without explicit reference to the dynamics connecting it to the community from which it has been elevated.

In practice, fitted interaction values may always be apparent interactions, because there will always be components that ecologists do not know about or have no data on—for example, larval and adult life stages will be lumped together, or decomposers or other components of the food web will be omitted. The critical question is whether the distinction between direct and apparent interactions matters. Do apparent interactions deviate significantly from direct observations of interactions or from ecologists' intuition about plausible interactions among populations? Ecologists would think that the protozoan populations in this example should be competitors because they share a food resource, but Vandermeer's study counters that idea. Can a more general conclusion be derived?

Apparent Interactions in an Eight-Population Food Web

One way to examine the importance of distinguishing apparent and direct interactions is through the following investigation or thought experiment involving model communities and ecologists who study them.[9] First, I take an all-knowing role and dictate the relationships among all the members of a model community. Then the hypothetical ecologists analyze data from the community and build a model of changes in the population sizes over time—their *trajectories*. But they collect data only for a subset of the full community, and the model they build refers only to the populations in that subset. Therefore, whatever interaction values they fit to the model will be apparent interactions that combine both the direct effects of interactions between modeled variables and the indirect effects mediated through the hidden variables. The ecologists in this thought experiment are skillful at curve fitting, so that the trajectories predicted by the subcommunity model mimic the actual ones as well as possible. Now, using my all-knowing position, I can compare the ecologists' model with the full model.

It is not quite that simple. The outcome of such a comparison would vary according to the form of the model used for the subcommunity, and so it would be difficult to make any generalizations about such comparisons. I can circumvent this limitation, however, if the full model has a feasible (i.e., positive) equilibrium and I restrict the comparison to the trajectories close to this equilibrium. Under these conditions, any form of the full model and the ecologists' model can be approximated well near equilibrium:

Model 2: Near-equilibrium linear form (NEL)
        Rate of change of population X =
        self-interaction within the Xs +
        sum of effects of each of the other populations on X

where the first term is a constant times the deviation of population X from its equilibrium value, and the interpopulation terms are constants times the deviations of the other populations from their equilibrium values.

Suppose I use model 2 to generate the population trajectories for the full community—strictly, the near-equilibrium approximation to the trajectories—and ecologists use model 2 to mimic the populations in the subcommunity only. I can calculate the exact values for the full model, and the ecologists can find values that give the best fit. When these values are compared, the ecologists' model necessarily mimics the actual trajectories imperfectly. However, despite their limited information, their version of model 2 can sometimes fit quite well.

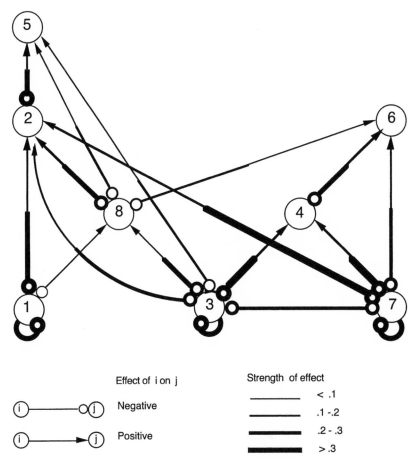

Effect of i on j

i———o(j) Negative

i———▶(j) Positive

Strength of effect

———— < .1

———— .1 -.2

▬▬▬ .2 - .3

▬▬▬ > .3

Figure 1.8 Eight-species model community, consisting of three plants, two herbivores, and three omnivores.

Consider the model community shown in figure 1.8, consisting of three plants, two herbivores, and three omnivores, and governed by a stable GLV model. (This model was generated through construction and turnover; see chapter 1, section A.) Suppose the ecologists restrict their attention to the subcommunity of the consumers—that is, they omit the plants (populations 1, 3, and 7). Figure 1.9A gives the trajectories for the five populations oscillating toward the equilibrium generated by the near-equilibrium approximation to the full model, and figure 1.9B gives the ecologists' best fit to this. The trajectories are clearly very similar.

Now compare the actual interaction values and the ecologists' apparent interactions (figs. 1.10A and 1.10B). The apparent interactions include

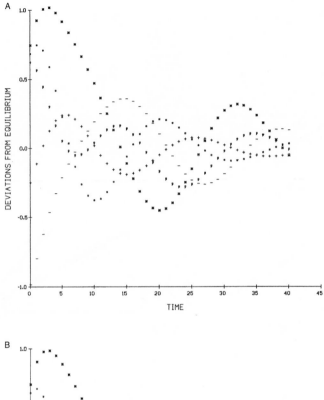

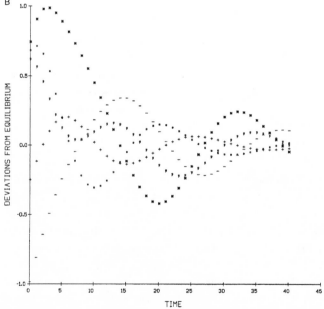

Figure 1.9 Changes in population sizes for the consumer subcommunity as the populations approach equilibrium, as determined by (A) the direct interaction values for the full community and (B) the apparent interaction values for the subcommunity alone.

self-inhibition by each consumer population; omnivore 6 and herbivore 4 preying on top consumer 5; omnivore 2 competing with both 4 and 6; and herbivore 4 both competing with its predator 6 and being the prey of another herbivore, 8. In addition, there are apparent interactions matching in sign the direct predator-prey interactions of the actual community. Whatever intuition one has about the effects of the hidden resources—the plant populations—it would not, surely, include a top consumer being a prey of lower consumers. Yet such are the apparent interactions the ecologists would find best mimicked the trajectories they observed.

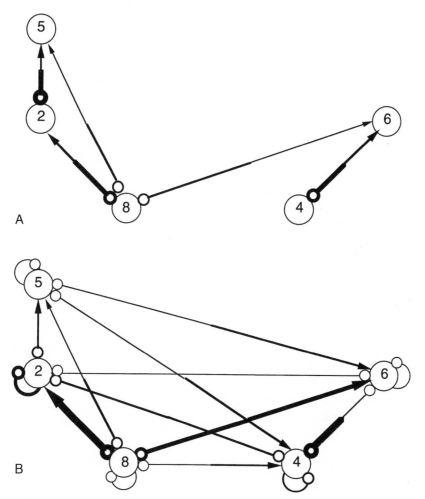

Figure 1.10 Interactions among populations in the consumer subcommunity: (A) direct; (B) apparent. Symbols are as in figure 1.8.

These results imply that understandings derived from analyzing simple subcommunities can be confounded by the dynamics of populations with which those subcommunities interact in naturally variable and complex ecological situations. Before exploring the implications of these results for modeling, the following two slightly technical points should be noted (Taylor 1985, 119–76):

1. The best-fitting apparent interaction values vary depending on the initial sizes of the population deviations from equilibrium. When the exercise above is repeated over a random sample of starting points, the average values and spread around these averages can be calculated. It turns out that some of the counterintuitive values disappear in the averages, but not all do.
2. When more variables are hidden, the fit between the ecologists' best model and the real trajectories becomes less satisfactory—as evident in figure 1.11. Furthermore, over a range of starting points for the trajectories, the spread of values around the average is relatively greater when more variables are hidden.

Making Sense of Apparent Interactions: A Dialogue

Apparent interactions have the following characteristics:

1. They can be counterintuitive; yet
2. they mimic well the trajectories of the populations.
3. The spread of the estimated interactions increases as the range of starting points is enlarged.
4. The fidelity of fit decreases as more variables are hidden.

Let me run through several different ways of making sense of these features by means of a dialogue with the ecologists.

Ecologists (E): Are you sure that the surprising values we obtained were not an artifact of the simplifying assumption that the populations are approaching an equilibrium?

My response (P): Vandermeer's "predator-prey" interactions indicate that counterintuitive apparent interactions are not restricted to near-equilibrium situations. Furthermore, by a continuity argument, the other characteristics (2–4) would hold even if you had derived the apparent inter-

TIME TRAJECTORIES OF DEVIATIONS FROM EQUILIBRIUM

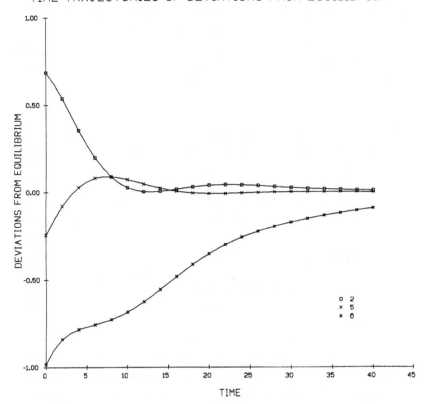

Figure 1.11 Changes in population sizes for the higher-consumer subcommunity (populations 2, 5, 6) as determined by the apparent interaction values for the subcommunity alone.

actions away from the neighborhood of an equilibrium. The only reason not to test you with an example away from equilibrium is that this brings in the additional problem of deciding the form of equations to use for the subcommunity.

E: It cannot be the case that all omitted variables will have a strong influence on the modeled relationships.

P: Conceded. If the timescales of the components of the subcommunity are much longer than those of the hidden variables (e.g., elephant vs. bacterial generation times), the hidden variables may equilibrate quickly and their effects on the subcommunity may be constant (Göbber and Seelig 1975). Or, if the interactions within the subcommunity are very strong, they may override the influence of the hidden variables.

Nevertheless, when you build models or formulate hypotheses to be tested, you ought to demonstrate—not simply assume—that modeled variables can be partitioned from hidden variables. Furthermore, you should not go out looking for cases that show partitionability—by virtue of interaction strengths or timescales—and then generalize this property to all subcommunities.

E: What information, then, do you think a well-fitting model can provide about actual ecological relationships?

P: The same method was used to derive the intuitive apparent interactions as the counterintuitive, so you should not try to give biological significance to the former and discount the latter. If Vandermeer's apparent predator-prey interactions do not require biological interpretation, then neither should his apparent competitive interactions. If you note that apparent interactions are sensitive to context and to starting points, then a good fit may simply mean that the hidden variables happened to remain within narrow bounds, or that you entertained only a limited range of starting points. Goodness of fit, achieved by Vandermeer's model and by your apparent interactions in the five-population consumer subcommunity, does not, therefore, indicate that a model represents actual ecological relationships.

E: What else would you want in addition to goodness of fit?

P: You should ask for evidence that your model contains the full community—or at least contains all of a subcommunity that is partitionable from other subcommunities. Furthermore, although it is rare in ecology to collect enough data to fit dynamic equations to them, this stipulation must apply equally well to the more common situation in which you merely assess the qualitative correctness of the model's predictions.

E: If the model fits well—whether or not it represents actual ecological relations—why not use it as a basis for predictions?

P: Some scientists judge a model by its predictive success; most would value a model that fits well over one that fits poorly. Note, however, that apparent interactions vary as the starting points of the trajectories vary. Suppose you have derived a well-fitting model. Even then, if you were to subsequently observe trajectories beginning with different starting points, the fitted parameter values would change, perhaps even qualitatively. This sensitivity would especially be the case if the original fit were for a narrow range of starting points or for one replicated starting point, which was the case in Vandermeer's experiments. Similarly, a well-fitting model might no longer fit so well if the composition of the hidden variables were to change.

Conversely, any predictions for the changed circumstances based on the original parameter values could be poor. You should, therefore, specify the range of circumstances in which the fit was derived, and recognize that beyond this range a well-fitting model is an uncertain basis for prediction.

E: There will always be hidden variables, except in experimentally controlled and isolated systems. On pragmatic grounds, why not eliminate the distinction between actual and apparent ecological interactions—that is, simply define the effect of one population on another to be the values we fit to a biologically sensible model? Most ecologists already do this in some situations—the concept of exploitative competition, for example, is explicitly one of apparent interaction: the shared resources are the hidden variables.

P: If you adopt this approach, then you should also note that apparent interactions might be counterintuitive. It would follow that, when formulating models and fitting them to the data, you should not incorporate constraints on your parameter values to make the models appear biological. In contrast, Vandermeer, for example, assumed that intrinsic growth terms had to be positive and self-interaction values had to be negative. (Paradoxically, he did not insist that the interpopulation interaction values—the pairwise interactions between the protozoans—had to be negative.) I would suggest allowing all parameters to be free, within the form of the model chosen, to take any value and sign. Furthermore, even if you keep the parameters of your models intuition-free, you could not expect the influence of hidden variables on these parameters to be constant over time or independent of the values of the hidden variables.

E: All models are simplifications or caricatures. We expect any model to depart from reality, and we expect these departures to guide us as we add biological detail to improve them. If we allowed concerns about hidden variables to inhibit the formulation of models, "there would be nothing to modify and we should get nowhere" (Hutchinson 1978, 40).

P: The rationalization that all models are caricatures is weak. You have seen that a well-fitting model may require counterintuitive parameter values in order to incorporate the effects of hidden variables. As a corollary, you should not translate your simple verbal models directly into mathematical terms. You cannot assume, for example, that exploiting a shared resource in a community is well represented by a negative-negative interaction in the subcommunity consisting only of those "exploitative competitors." Correspondingly, theory based on simple models, such as competitive exclusion demonstrated on two-population phase diagrams, becomes problematic. In short, there is a hidden complexity to simple models.

Furthermore, unless you know that the model contains the full community, you cannot argue that lack of fit or counterintuitive parameter values signify that some biological feature is missing from the model. For example, the stated purpose of Vandermeer's study was to detect "higher-order" interactions. By modeling only a subcommunity, however, he could not resolve issues about the actual biology. Additional terms may have improved the fit of Vandermeer's model, but it is not warranted to describe this as improving the model by the addition of biological detail.

E: I wonder if this conclusion was an artifact of your use of the sparsely parameterized GLV model.

P: I'm not sure what grounds you have for being so skeptical. However, it would be possible—albeit an enormous project—to repeat the thought experiment to check whether hidden variables confound specific models that include a large number of parameters—for example, parameters related to the behavior and physiology of individual organisms, or to the flows of nutrients through ecosystem compartments. In the meantime, I recommend that ecologists revisit debates about the value of models based on deterministic relations, such as consistent interpopulation interactions. When the fit to observations of such models was not significantly better than that of corresponding "null" models (Strong et al. 1984), the existence of the deterministic relations was called into doubt. A plausible alternative hypothesis, however, is that model specification is incomplete and the relations exist, but are confounded by hidden variables (Schaffer 1981).

E: Are you saying that ecological models need to include every bit of detail or else they are, strictly speaking, biologically irrelevant?

P: No, not every detail, but at least include variables having dynamics of their own.

E: Suppose we controlled those variables, and they became merely parameters, then we could get more biological mileage out of our models of subcommunities, right? We could, for example, try to redo Vandermeer's experiment keeping the bacterial populations more or less constant.

P: In principle, yes. Simple mathematical models have sometimes proved effective for laboratory microcosms (Williams 1972)—but not always (Mertz and McCauley 1980). This strategy would, however, sidestep the primary issue of models of non-laboratory ecological situations.

E: Sidestep? It is a standard scientific strategy to learn about the functioning of variables in a controlled situation and then to use that knowledge
to help understand their functioning in a larger context (Bechtel and
Richardson 1993).

P: Yes, it is a strategy, but an ambiguous one for ecological research.
Some scientists and philosophers of science like to see it as a way to
expose or localize the mechanisms used by organisms, and, at least in
principle, to derive from the localized mechanisms an account of how the
organisms function in uncontrolled situations. In contrast, you can also
think of the strategy as a heuristic use of redescriptions. What do I mean
by this? Suppose a model draws your attention to a conjunction of certain
factors and, for the purposes of making predictions or in further investigations, you focus on these factors as if they governed the phenomenon
under consideration. You are using the model heuristically if it guides
your work in this way, and you also keep in mind that the model will
break down when applied too far out of the domain in which it was
derived. In this spirit, when a model fits the quantitative data, but you
think it captures apparent interactions—not the actual causal relations—
you could think of the model as a redescription of the data that you use
heuristically.

E: There seems to be a circularity in your strategy. To specify how far
out of the domain is too far—in our case, what changes in the hidden variables and starting points are great enough to require qualitative changes in
the model—we would need to know more about the dynamics of the full
community. And if we knew that, we would not have to restrict ourselves
to a model of the subcommunity.

P: I agree. Suppose, then, that you don't wait until you can write
down a model for the full community, but use redescriptions as a basis
for new hypotheses, experiments, and generalizations. That's OK, but at
the same time you need a way of questioning the scope of the model's
application. Any model used to guide your thinking may turn out to
have misguided you, and you don't want to go a long time without finding out.

E: Instead of playing around with fallible models, could we avoid the
problem by including all the variables, or, at least, omitting fewer of them?
It seems to us that systems ecologists are less likely to omit variables
from their models because they trace the flow of nutrients and energy
through an entire ecological system, especially through the decomposer
components.

P: The full system consists not only of variables, but also of their dynamic interrelations. Systems ecology, in my opinion, has been too ready to translate measurements of covarying variables into equations without elucidating the biological dynamics (see chapter 3). Instead of trying to be all-inclusive, I think we need to find ways to use models heuristically, all the time checking that they are not applied too far out of the domain in which they were derived.

E: In this respect there may be something to be learned from Holling's (1978) idea of "Adaptive Environmental Management." Holling promotes use of multiple models and their ongoing revision in recognition that any ecological situation is a moving target—not the least because management practices produce continuing changes.

P: I agree. Ongoing assessment might allow us not only to correct for the confounding effects of dynamics not in the model, but also to take account of turnover in the components of the modeled system (chapter 1, section A). But it is an open question how knowledge production in ecology can be reorganized so that models are used heuristically and ongoing assessment and revision is built in—even more so if ecology accommodates management and exploitation by humans. It is a question for us to address together. In the meantime, let me summarize what apparent interactions show you about the strategy of searching for basic principles about the interactions of a subset of the ecological community and building up a picture of ecological complexity based on such principles:

Unless you know that the full community has been specified, a model is primarily a redescription of the particular observations that does not provide, through its fit or its lack of fit, sure or general insight about actual ecological relationships. If this conclusion seems too extreme to accept, it should at least challenge you to clarify your reasons—which may be sociological as well as scientific—for building models that refer to only a few populations when those populations are embedded in a dynamic ecological context and responding to consumers and resources that are unevenly distributed across place and time. Simple models mostly hide their complexity.

E: Hold on. Your points must also apply to other fields, such as economics and social theory, when their models and conceptual frameworks omit explicit reference to variables that have dynamics of their own.

P: I think so. Let's keep this in mind.

▲

Two conclusions had emerged from my exploration of the consequences of ecologists defining boundaries between the outside and inside of a system when they attempt to account for ecological organization:

1. *Ecologists interested in explaining the persistence of complex communities need to examine not only the current configuration or "morphology" of that complexity, but also its construction over time—its contingent history of becoming structured and its ongoing restructuring in a wider spatial context or "landscape."*

2. *Ecologists need to recognize that principles derived from analyzing simple subcommunities can be confounded by the dynamics of populations with which those subcommunities interact in naturally variable and complex ecological situations.*

It seemed that construction over time together with embeddedness in a dynamic context, which links the system that is the focus of research with the backgrounded processes, have potentially profound implications for knowledge-making: Theorists should not assume that ecological complexity can be partitioned into communities or systems that have clearly defined boundaries, coherent internal dynamics, and simply mediated relations with their external context.

Around this time—the mid-1980s—I became aware of the work of the anthropologist Eric Wolf, which primed me to look in areas other than ecology for ways to think about problematic boundaries: "Societies emerge as changing alignments of social groups, segments, and classes, without either fixed boundaries or stable internal constitutions." If anthropologists observe "transgenerational continuity, institutional stability, and normative consensus," they should seek "to understand such characteristics historically, to note the conditions for their emergence, maintenance and abrogation. Rather than thinking of social alignments as self-determining ... we need ... to visualize them in their multiple external connections" (Wolf 1982, 387). In other words, whenever theory has built on the dynamic unity and coherence of structures or units—in Wolf's case, societies or cultures—researchers could invert this and consider what would follow if those units were to be explained as contingent outcomes of intersections among processes that implicate or span a range of spatial and temporal scales. As will emerge later (chapter 5, section C; chapter 6, section A), socio-environmental studies proved to be a more fertile field than ecology proper for me to elaborate on Wolf's conceptual inversion and paint a picture of intersecting processes (see, e.g., Little 1987; Peet and Watts 1996a; Taylor and García-Barrios 1995). Nevertheless, even within ecology proper, my inquiry into the relations between models and the reality to which they refer (chapter 2) would lead me again to the core conceptual issue of embeddedness in a wider context ...

▲

2

Open Sites in Model Building

Boundaries become problematic when they discount history, embeddedness in a spatial context, or the dynamics of variables not explicitly included in the models: such was the conclusion I drew from the modeling presented in chapter 1. This perspective is especially challenging for mathematical modelers because the assumption of a fixed, delimited set of components is almost required for formulating and analyzing a mathematical model. Recognition of this fundamental limitation of ecological modeling started me thinking about the need for some ongoing process of assessment, reformulation, and reassessment. I was primed to notice analogous problems when I ventured into the interpretation of science and other fields (see chapters 3–6).

I did not, however, abandon modeling and adopt the view that ecology should consist of particular case histories. Models had proved valuable; my understanding of problematic boundaries had been derived through theorizing that centered on models. With regard to the construction of complexity, I had followed the tradition in which simple models are used to seek qualitative and general insights (chapter 1, section A). To understand the sensitivity of the principles ecologists derive about subcommunities to the dynamics of the other populations in the community, I had explored a model world in which I knew the complete dynamics, but the hypothetical ecologists analyzed data from the subcommunities only (chapter 1, section B). These two modeling exercises were complemented by my earlier experience in agricultural research. When I had to extract patterns from data on the complexity of interactions between plant varieties and field conditions in large crop trials, I followed the lead of certain vegetation ecologists. These researchers had used models to generate artificial data, which allowed them to examine the sensitivity of their multivariate data analysis techniques to built-in assumptions about the nature of the causes that the techniques are intended to expose (see note 10).

My awareness of the tension between the productive potential of model-based theorizing and its limitations led me to try to make sense of the positions that philosophically minded ecologists were staking out during the 1980s. What do models model? This is the subject of chapter 2 ...

◆ ◆ ◆

Should modelers favor generality and realism, realism and precision, or neither combination?

In 1966 Richard Levins sketched a strategy of model building in ecology and population genetics that favored sacrificing "precision to realism and generality." Models, he said, should be seen as necessarily "false, incomplete [and] inadequate," but productive of qualitative and general insights. Discrepancies between a model and observations imply the need for additional biological postulates and, together with the qualitative insights, generate interesting questions to investigate. Eventually a model becomes "outgrown when the live issues are not any longer those for which it was designed" (Levins 1966). In practice, this strategy applied to community ecological models simple enough to be analyzed mathematically, not to highly parameterized systems ecological models that required computer simulation. This emphasis was taken up in mathematical ecology, a field that grew markedly in the 1970s. (Mathematical ecologists often left ambiguous, however, whether they saw their models as idealized representations of ecological reality—e.g., the "perfect crystals" of May 1973—or as heuristics to help guide theoretical inquiry.) In the subsequent decades mathematical modeling in ecology has proliferated and grown increasingly sophisticated.

In the early 1980s, however, a strong reaction developed against ecological theory drawn from simple, general models. Such general principles could be rejected if alternative "null" or "random" models fitted the data equally well. Rejection of previously proposed models led, in turn, to an emphasis on experimental testing of specific hypotheses about particular situations (Simberloff 1980, 1982; Strong et al. 1984). In short, according to skeptical and particularist ecologists, the previous models were not realistic, and general models were not likely to be found in ecology. Realism and precision would be possible for models of particular situations, not for big questions such as accounting for ecological complexity. In later assessments of Levins's article, some philosophically minded critics have also focused on his classification of models as general, realistic, or precise, and have argued against a necessary trade-off among these three qualities (Palladino 1991; Orzack and Sober 1993).

There have also been many criticisms of the particularist reaction and defenses of a strategy of model building along the lines Levins sketched (May 1973, 1984; Hutchinson 1978; Levin 1980; Levins and Lewontin 1985, 132–60; Roughgarden 1983; Hall and DeAngelis 1985; Caswell 1988; Levins 1993). A common theme has been the value of models for stimulating the development of evolutionary and ecological theory. Chapter 1 makes evident my sympathies with a stimulatory or *exploratory* role for models in ecology. However, I want to go further than the authors just listed and expose some more profound implications that the exploratory approach has for thinking about science. In the process of analyzing what modelers are doing when they model, a different view of Levins's strategy and the generality-realism-precision trade-off will emerge.

By way of preliminaries: I take models to differ from theories in that theories are a "fabric" of one or more models. Only some models—those at the edge of the fabric—are exposed for empirical scrutiny. Other models woven into the theoretical fabric may be left inexplicit. Some models take the form of verbal or pictorial *representations* or material devices, but I focus here on mathematical equations of dynamic relations and on models that could be rendered readily into mathematical form. (For some contrasting formulations of modeling in ecology and related fields, see Harvey 1969, 162–76; Oreskes, Shrader-Frechette, and Belitz 1994; Peters 1991.)

Three Levels of Model Building

Levins's commitment to social change has led to extensive critical work on social shaping of scientific knowledge (Levins and Lewontin 1985; Taylor 2000c). Nevertheless, his account of modeling retains reality as the benchmark. All models are "incomplete"—they cannot possibly include all aspects of reality. Nevertheless, they can provide insights about the real world. Discrepancies between a model and observations are taken as grounds for adding *biological* detail to the model with the aim of tightening the correspondence. Models may be false, but they are "means to truer theories" (Wimsatt 1987).

Let me refine this picture of models as representations of reality. (I will return later to questions of social influences on science.) All models necessarily simplify reality, but they are not all designed and applied according to the same standard for correspondence with observations. Some models are evaluated by a quantitative analysis of correspondence between patterns in the data and predictions of the model. Other models, or the same models in different situations, gain acceptance according to the plausibility of their assumptions and predictions.

Furthermore, there are two faces to any model. On one face there are some *distinguishing features*. Levins's loop analysis, for example, models a system of interrelated variables, such as a community of interacting populations, near equilibrium (Puccia and Levins 1985). The near-equilibrium assumption means that the equations for the rate of change of each variable can be linearized—that is, simplified to additive combinations of population sizes. Algebraic rearrangements of these equations then allow the modeler to predict the effect of one population on another population when the conditions, and thus the equilibrium values, change slightly: do the values go up or down together, in different directions, or remain unchanged? This is the obvious face of loop analysis. On the other face there are *accessory conditions* that are built in and assumed when analyzing a model but are not its distinguishing features. For example, loop analysis assumes that the system is at an equilibrium or can regain it rapidly after any disturbance, that the environment of the populations is constant in space and time (Williams 1972), and that the modeled populations are effectively independent from the unmodeled or hidden variables (chapter 1, section B). Accessory conditions are often overlooked or difficult to establish. This is especially true for partitionability, a condition that requires the unmodeled context to enter only through the constant parameters of a model (chapter 1, section B).

Figure 2.1 presents a schema of model *confirmation*, which incorporates the model's accessory conditions as well as the distinguishing features that characterize a dynamic or causal model. Three broad levels of correspondence between a model and observations can be distinguished; these levels are summarized in table 2.1 (Taylor 1989b; see Lloyd 1987 for a related discussion of theory confirmation):

1. A *framework* consists of one or more propositions or themes that guide our inquiry by focusing attention on certain biological situations and processes—for example, in loop analysis, population growth and limitations on or facilitations of that growth. The conceptual implications of the propositions can be explored without examining how closely they

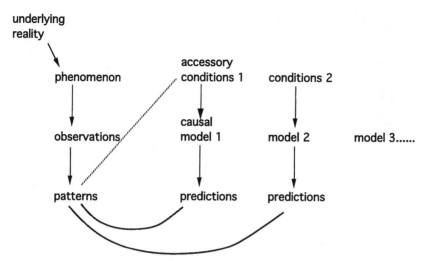

Figure 2.1 Schema of the confirmation of scientific models. Arrows indicate the sequence of steps in modeling; the dotted line and the gray curves indicate the two aspects of the analysis of correspondence.

correspond to detailed observations, and this *exploratory modeling* may allow ecologists to formulate new questions, new terms, or new models. When the propositions are formulated in mathematical equations, the framework can be explored systematically as a *mathematical* system. For instance, it becomes possible to ask how the system's behavior changes as its parameters change or become variables, as time lags are added, and so on. (The term *heuristic* can be used in the sense that I use *framework* or *theme*, but *heuristic* also has connotations of a "rule of thumb," which raises different issues.)

2. A *redescription* is a summary of observations in the form of a model whose parameter values have been estimated from the data. It permits prediction or extrapolation on the basis that observed patterns might continue into the future or extend to other situations.

3. A *generative representation* is a model that not only fits the observations, but for which evidence for its accessory conditions exists *independently of that fit*. If both kinds of correspondence have been demonstrated, ecologists are justified in acting as if the model represented the dynamic biological relationships that generated the observations. Moreover, because the model captures the necessary and sufficient conditions to explain the phenomenon observed, confident predictions can be made for situations not yet observed.

Table 2.1 Levels of correspondence between a model and observations

Level of correspondence	Fit	Accessory conditions
1. Framework	—	—
2. Redescription	✓	—
3. Generative representation	✓	✓

✓ = quantitative analysis of correspondence with data
— = correspondence judged to be plausible or not examined

By making the preceding distinctions, modelers can clarify the level at which they accept a model or intend it to be used. Such clarification might counter the tendency of field ecologists to judge models harshly—as if only the last level, that of generative representations, were significant. Different applications are appropriate to each level of correspondence, and models should not be judged equally as representations of reality. On the other hand, models should not be excused indiscriminately on the grounds that the questioner is asking too much of the model. By stating what level of correspondence is intended, modelers can clarify how seriously their models are to be taken.

Nevertheless, even if modelers clarified the level of correspondence of their ecological models, the matter would be unlikely to rest at that. Two factors ensure that much free play would remain in the evaluation of correspondence:

1. The correspondence of models and observations is *relative*, varying according to the degree of fit and the strictness with which accessory conditions have been established. Disagreements about acceptable fit abound; sometimes assumptions that are self-evidently correct to one person can even be implausible to another. For example, a common assumption made by most researchers studying microevolution or microeconomics is that individuals are self-interested and maximize utility (but see Glimcher 2003). Some researchers, however, find it difficult to conceive ways that individuals could put utility maximization into practice when their lives depend on a diversity of resources and involve them in processes that span a range of spatial and temporal scales (see chapters 4–6, especially chapter 5, section C).

2. Correspondence is also *provisional*, contingent on the variety of circumstances in which the fit and the accessory conditions have been established,

on the stated level of resolution, and on the range of accessory conditions exposed for scrutiny. At some level of resolution, most accessory conditions will cease to hold. For example, organisms that modelers have abstracted as identical units, in order to subsume them into a single variable in the equations of their models, are usually genetically heterogeneous (Lomnicki 1980).

The relative and provisional nature of correspondence means that the three levels of correspondence become ideal types whose boundaries are, in practice, indistinct. The point of the three-part taxonomy, however, is not to provide an unambiguous classification, but to facilitate discussion of any model's strengths and limitations with a view to revising it. In this spirit, the levels of correspondence might be better thought of as levels of modelers' *acceptance* of models. By mobilizing the free play involved in evaluating correspondence, ecologists can reconsider—or can be pressed by others to reconsider—their acceptance of a model. The outcome of such reconsideration is more likely to be revision of the model than simply its rejection or acceptance. Revision, moreover, is not necessarily directed at tightening the fit of a model; it may run a gamut from attempting to expand acceptance of the model to attempting to disturb its acceptance. That is, in practice, a modeler might make any of a number of moves:

1. *elevate* the model's status (e.g., convert a framework into a generative representation)
2. *generalize* (e.g., claim an expanded domain of application for) the model
3. accept the model and *shift focus* (e.g., to other models in the theory)
4. *refine* the model (i.e., attempt to improve the fit by adjusting parameter values or adding details)
5. *qualify* the model (i.e., give more consideration to accessory conditions)
6. *disturb acceptance* of the model (i.e., search for circumstances in which confirmation breaks down—in Popperian terms, test a risky prediction)
7. *reconfigure* the old model into one with new variables and relationships (which is Levins's favored move: Taylor 2000c).

Revising Models and Generating Theory

It might be possible to assimilate the seven directions for revising models to a perspective that centers on hypothesis testing. Generalizing, for example, could be viewed as merely producing a new hypothesis to be tested. However, a methodology of science that emphasizes hypothesis testing can say little about the generation of theory. What Levins advocates is, in effect, theory generation through exploratory modeling. By favoring generality and realism at the expense of precision, he keeps the focus away from three alternatives: establishing generative representations, framing every new idea as a hypothesis to be directly tested, and building detailed models of specific situations.

It would help ecologists who are interested in generating theory if they distinguished three levels of correspondence and different directions of model revision. This clarification would remind them that not only the distinguishing feature of the model, but also the accessory conditions, might be questioned. Furthermore, framing any new idea as a hypothesis to be tested is not necessarily the quickest route to better generative representations. Suppose, to choose a simple example, one wanted to reevaluate the logistic equation as a generative representation of the population growth of a single species. (Growth that follows the logistic equation begins exponentially but slows down to zero as the population approaches a limiting level—the carrying capacity; see Vandermeer 1990, 6ff; Williams 1972.) Instead of testing a prediction of the model, it could be illuminating to move back to the level where the logistic equation is a framework used for exploratory modeling. The effects of allowing the model population to be genetically heterogeneous and spatially distributed, for example, or of including the population's resource and other hidden variables in a new model, could be examined. Out of such explorations might emerge ideas about the conditions under which the logistic equation could work as a generative representation.

The value of exploratory modeling in shifting theory onto new ground is exemplified in chapter 1, section A. Early mathematical exploration of the stability-complexity relationship indicated that complexity detracts from community stability (May 1972) and stimulated a search for "the devious strategies which make for stability in enduring natural systems" (May 1973, 174). The theoretical and empirical research that followed—into spatial heterogeneity, disturbance, and so on—has been of value, despite subsequent indications that there may not, in fact, be any clear stability-complexity relationship (Goodman 1975; DeAngelis and Waterhouse 1987). The complexity-instability result also led some modelers to explore the addition and elimination of components. The construction of model systems over time allows complexity to persist even when any particular system is transient. From this exploration arises the nontrivial theme that investigations of ecological complexity should include the historical development of that complexity and not simply analysis of its current configuration.

The virtues of exploratory modeling for generating theory need, however, to be qualified in several ways. The attribution of generality and realism to Levins's strategy is problematic. Strictly speaking, without a quantitative analysis of correspondence, the insights from exploratory modeling are insights about a mathematical system. Their relevance to biology is yet to be established; truth or falsity is thus a moot issue (contra a strict reading of Levins 1966 or Wimsatt 1987). Instead, mathematical results can be thought of as themes or a framework that guide or stimulate subsequent research.

The use of exploratory modeling results in an uneasy tension between the need for mathematical tractability and the demand that the exploratory model eventually be more strictly evaluated against observations. A framework may turn out to *mis*guide those who rely on it. The categories built into exploratory models are often chosen with an eye to mathematical tractability, which may obscure profound issues of biology. For example— venturing outside ecology here—since their origin during the 1920s, population genetic models have been built up from ideal genotypes that possess different fitness values. Within this framework it is difficult to incorporate the construction of characters during ontogeny or to examine the evolutionary significance of such construction (Oyama 1985; Oyama, Griffiths, and Gray 2001).

In this light, my resistance to an emphasis on hypothesis testing is not an endorsement of general propositions that can be expressed in simple models. The once popular analogy of simple ecological models with the perfect crystals (May 1973, 11) or ideal gases of physics and chemistry is misleading. In physics it is possible to characterize the conditions, such as low pressure, in which the ideal is approximated, then go on to produce those conditions by experimental manipulation and observe whether the actual behavior approaches that of the modeled ideal gas. In ecology, it is difficult to characterize in biological and noncircular terms all the required conditions, especially partitionability (chapter 1, section B). Attention to such accessory conditions exposes the hidden complexity in simple models. The themes derived from simple mathematical models may be valuable for generating theory, but are not shortcuts to ecological generality.

Another problem in exploratory modeling is subtler. The support of such modeling by authors such as Levin (1980), Hall and DeAngelis (1985), and Caswell (1988) rests in part on an unstated implication that, if the themes from different exploratory models could be combined, they would yield an understanding of ecological phenomena unachievable through the construction of all-encompassing systems models. The means, however, of weaving together or synthesizing themes has yet to be articulated by advocates of exploratory modeling. (In a related context, Hilborn and Stearns 1982 provide some warnings about problems that must be addressed in extending single-factor results to situations in which multiple causes interact.) If, as I suspect, no one ecologist can juggle more than a few themes, new approaches are needed that bring ecologists, employing various different themes, into sustained interaction (see chapter 5, section B; chapter 6, section C; and the epilogue, section B).

Exploratory modeling is not necessarily the only way, nor the best way, to generate new models and theory. An insight such as the need to examine the historical development of ecological complexity could be raised by, say, discussion and contemplation of models at a thematic level. Such discussion could arise from what Haila (1988) calls the "many faces of data"— the left hand side of figure 2.1—which include initial category-generating generalizations from observations, comparisons, analytic redescriptions such as those produced by multivariate data analysis (Austin 1999), and field experiments designed to allow multifactorial statistical analysis (Underwood 1997). Yet even when ecologists emphasize data analysis, exploratory modeling can help them illuminate the problems of reliably inferring dynamics from patterns and the sensitivity of such patterns to the methods used to extract them from data.[10]

The qualifications about exploratory modeling expressed in the preceding paragraphs, however, reinforce rather than undermine the general point that revision at all levels can stimulate ecologists to generate different models. The larger the range of competing models, the more active the theorizing. An emphasis on revising models, which builds on the schema of model justification or acceptance, puts me in a position to raise an additional set of important issues.

Open Sites

Revision cannot be the ultimate source of competing models. How in the first place would it occur to ecologists to explore, for example, the possible effects of heterogeneity within a population or to model each individual separately (Huston, DeAngelis, and Post 1988; DeAngelis and Gross 1992)? Indeed, have ecologists overlooked other ways of questioning any given model? The obvious response is to acknowledge that at some point ecologists have to borrow from other fields and other situations in order to invent new models. Yet even when scientists actively revise their models, it is not inevitable that they will invent new ones—categories can remain plausible for long periods of time with competing models scarcely considered. At this and several other points in the process of model justification that I have sketched above, modelers have to make important decisions in which something other than experiments, observations, and comparisons must come into play. Let me characterize six of these points:

1. Strictness of confirmation. Nature does not tell scientists what degree of fit between models and observations is acceptable or necessary for good science, nor how strictly accessory conditions should be established. Instead, groups of scientists make these decisions, invoking one or more of the following, not necessarily consistent, considerations:

    a.   previous experience or standard practice in the field
    b.   requirements for technical control over the system
    c.   comparison with a null model (i.e., one without the distinguishing feature)
    d.   comparisons with a range of competing models (see item 3 below)
    e.   openness to periods of exploratory modeling, during which analysis of correspondence is looser

2. Acceptance versus disturbance of confirmation. Given that correspondence is provisional, scientists have to decide in which direction they will move—toward accepting a model or toward disturbing its current degree of confirmation.

3. Range of competing models. The range of models available to be compared may not be exhaustive because there is nothing inevitable about the creation of alternative models. Even when many models are available, scientists choose whether to subject all or only some of them to comparison.

4. Technical considerations. The range of competing models may be constrained by the abilities of mathematicians and their computers to elucidate the behavior of different models, to calculate equilibria and stability, and so on. Although new developments in science may eventually loosen some technical constraints, there are modelers—Levins included—who prefer models with dynamics that can be analyzed and understood relatively easily. They dislike models composed of a complexity of different kinds of equations and parameters, holding that the results, necessarily produced by computer simulations, are difficult to explain in terms of the constituent dynamics. Levins argued, in this spirit, that the elaborate sets of equations in systems ecological models are unlikely to provide insights that can be applied or adapted to other situations. (Palladino [1991] reviews counterclaims.) At the same time, technical considerations have to be addressed if themes from different exploratory models are to be synthesized. In short, technical constraints mediated by stylistic concerns constitute a fourth point at which modelers' decisions are not driven by experiments, observations, and comparisons.

In a complementary way, technical considerations mediated by statistical conventions constrain the extraction of patterns from data. Conventions of statistical interpretation of explanatory claims are unavoidable because models of causes are not related in straightforward ways to the models that lie behind statistical techniques (note 10; see also Lewontin 1974). The statistical techniques scientists have available to use on their data also influence their design of experimental trials, as discussed under points 5 and 6 below (Underwood 1997).

5. Construction of phenomena. Few sciences confine themselves to the raw phenomena of nature. Instead, they establish conditions in which nature is placed under greater control or is made more reproducible, often with the result of creating situations that never occur otherwise. In ecology, field locations are selected, species or varieties transplanted into defined habitats, microcosms established in laboratories, multifactor field experiments conducted, and so on. (Phenomena first constructed by scientists and others are sometimes perpetuated so as to become a long-lived part of nature; for example, hedgerows, forest plantations, waste treatment plants.) Even though the construction of phenomena is often held to be a way to systematically expose the underlying reality of nature, nature does not tell scientists how they should be constructed.

6. Construction of observations. The translation of phenomena into observations and data requires choices of categories, sampling frames, the spatial and temporal extent of observations, and recording equipment. Technical considerations (see point 4 above) are involved in the design of equipment and the experimental construction of observations, and consequently, dictate which models can be supported empirically. Social considerations, such as what funding scientists can secure, also influence the construction of observations.

◆ ◆ ◆

The existence of these six points shows that ecologists cannot establish a relationship of models to ecological phenomena without making decisions in which something other than experiments, observations, and comparisons must come into play. This must be so even if these decisions involve conventions that are taken for granted by the particular group of scientists. In any case, conventions are the result of past negotiations among scientists, so in general, the six points can be seen as *sites* where modeling is *open* to negotiation and wider influences (fig. 2.2). (And what holds for models must also hold for theories, which consist of interconnected models.)

There is, of course, a way such decisions can be made to fit a more conventional focus on models and the phenomena to which they ostensibly refer: Decisions are made, but those that persist and become conventions have shown their value through the effectiveness of subsequent research and applications based on them. This claim (which philosophers describe as *realism*) discounts the scientists' decisions as theoretically superfluous, or at least they become so by the time the scientific community has reached a consensus or standardized practice (Taylor 1990).

This argument from persistence, however, downplays the *process* of modeling, and instead revolves around the correspondence of the *product* of modeling to natural reality. Yet all products emerge from a process. The process may conceivably be an evolutionary one, in which those ideas that correspond best to reality survive through experimental tests and disputes over correct interpretations. However, such schemes for the evolution of ideas or conventions leave unclear what scientists *do*. Unlike genetic

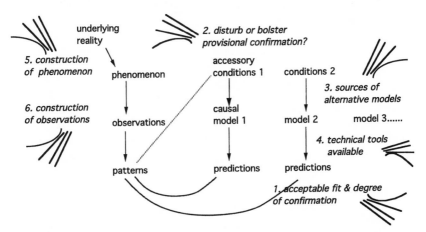

Figure 2.2 Open sites (where researchers' decisions depend on factors other than experiments, observations, and comparisons) superimposed on the schema of the confirmation of scientific models from figure 2.1. The clusters of lines entering each site indicate the potential diversity of influences on decisions.

mutations, the ways in which scientists vary their ideas or their decisions at the open sites are not random. When the processes through which scientists bring about the *survival of the most effective* are specified, evolutionary schemes tend to collapse to a tautology in which the most effective are equated to those persisting at any given time. The implied strategy of modeling—a distinctly non-Levinsian one—is simply to stay close to what is currently accepted.

Instead of accepting the argument from persistence, I suggest that ecologists acknowledge that modeling and theory building operate within webs of social and technical decisions. Of course, such a picture of embeddedness is abstract, for it does not specify the reactions that take place at the open sites—the negotiations and other considerations that influence the decisions. To use Levins's terms, the expanded account of model building is general and realistic, but not precise. It does, however, point to new questions to ask, new terms to employ, and different models to construct. Recall that exploratory modeling appeals to those who think that ecological theorizing can be stifled by an emphasis on hypothesis testing. Similarly, once ecologists recognize that concentrating on the validity of models and assumptions leads them to neglect the reactions at the open sites, they should see that a wider examination is needed—one in which considerations other than explicit analysis of models' correspondence with evidence must come into play. In short, exploration of how ecologists model ordered complexity is incomplete without attention to "modeling" of another kind; namely, of ecologists in the contexts in which they work. This is the subject of the chapters ahead.

▲

My approach to philosophy of modeling depended on first classifying different things that ecologists do when they build models to represent ecological complexity. There was an important place in this classification for exploration of concepts and generation of theory; ecologists discount this dimension of science when they emphasize testing specific hypotheses about particular situations. I was also able to identify several open sites, in which considerations other than analysis of a model's correspondence with evidence must come into play. In other words, knowledge-making must always extend beyond the dialogue between models and evidence. By analogy with a theme from my account of apparent interactions in ecology (chapter 1, section B), the dynamics of the wider influences on knowledge-making may confound any philosophical analysis of modeling and theory building that leaves those influences hidden. My recognition that dialogue is embedded in a dynamic social context opened up the larger project that is the subject of parts II and III ...

▲

Part II

Interpreting Ecological Modelers in Their
Complex Social Context

The motivation for the modeling efforts described in part I was to account for order in eco-logical complexity. However, after sites are identified where considerations other than explicit analysis of a model's correspondence with evidence must come into play (chapter 2), a wider exploration of ecological theorizing is opened up. What factors influence the deci-sions that ecologists make about which questions to put to nature, categories to use, obser-vations to construct, analyses to perform, degree of confirmation to require, and ways to revise models? Is the effect of these factors on ecological science merely idiosyncratic and transient, or are there systematic patterns? Could awareness and discussion among ecolo-gists of any systematic effects influence their subsequent science in productive ways? In particular, could such a wider contextualization of ecological theorizing help ecologists address the challenge of making sense of ecological complexity that involves ongoing restructuring and embeddedness (chapter 1)?

I had the opportunity to examine these new questions during research fellowships in two interdisciplinary programs. The first step I took was to relate my interests to the existing approaches in the interpretation of science. Some scientists and philosophers concerned with scientific method identified the different theoretical heuristics applied in science and com-pared their effectiveness in establishing knowledge (Bechtel and Richardson 1993). This approach focused on the dialogue between theories and the evidence about reality to which they refer. Historians and sociologists of science, on the other hand, tended to focus on inter-actions among members of scientific communities during disputes and dialogue around methods, observations, and conclusions. The resulting interpretations of science invoked a wide range of social factors—mentoring and favoritism, competition for prestige and public-ity, government or corporate funding decisions, gender relations and class interests, and so on.

It was clear that scientists and philosophers of science tended to assume—as do some historians and sociologists of science—that scientists' contributions to the dialogue between models and the evidence they refer to can be separated from their dialogue with other scientists to establish what counts as knowledge. This separation of the referentiality of science from its sociality might occur in a number of ways. The effect of decisions made at what I call the open sites (chapter 2) leads some scientists to tackle anomalies that others have dismissed as negligible and thus ensure that science progresses. In any community of scientists, disputes are resolved when one scientist's biases are countered by those of others; thus science self-corrects. Social influences, such as research funding, merely inhibit or accelerate improvement in scientific knowledge. In short, although science is a social endeavor, its referentiality still determines what counts as knowledge—if not immediately, at least in the long run.

I was more interested, however, in a deeper sense of sociality that makes it harder to keep sociality and referentiality separate. All scientists engage in various arenas of social activity—they build careers and institutions, use and transform language, facilitate policy formulation, and so on. This context means that scientists select problems, define categories, collect data, and present findings not only to develop models of their subject matter, but also to secure the support of colleagues, collaborators, and institutions and enable others to act on their conclusions. I realized that this might happen in idiosyncratic ways, but it was also possible that the simultaneous pursuit of referentiality and social support could sometimes lead to systematic and enduring effects on the content of scientific knowledge. I wanted to attempt to demonstrate and analyze such effects—the subject of part II—for two reasons: First, it would be harder to dismiss as "insignificant in practice" the conclusion from chapter 2 that considerations other than explicit analysis of a model's correspondence with evidence must come into play. Second, it would open up the possibility that ecologists who theorize about ecological complexity—or researchers more generally—might be encouraged to use awareness of such effects to modify their own work. This possibility is taken up in part III …

3

Metaphors and Allegory in the Origins
of Systems Ecology

I began to explore the effects of the sociality of science on its content in the field of systems ecology; in particular, in the work of H. T. Odum, a pioneer in systems ecology in the United States. Although this field emphasizes nutrient and energy flows, which were not often examined by the modelers in the stability-complexity debate (chapter 1 and note 5) stability was a central concern of Odum. Moreover, the shift to a field that explicitly considered entire complex systems allowed me to focus on the implications of partitioning complexity into systems assumed to have clearly defined boundaries, coherent internal dynamics, and simply mediated relations with their external context.

As an interpreter of science, I looked for correlations between scientific ideas and the scientist's social context and personal history. The scientist in me, however, was interested not only in correlations, but also in the mechanism or dynamics producing them. As I interpreted Odum's work, I also sought a plausible model of the dynamic relationships among his social and scientific ideas and practices ...

◆ ◆ ◆

A. Social-Personal-Scientific Correlations in the Work of H. T. Odum: A First Reading

How do scientists produce the *social-personal-scientific* correlations that historians detect?

When historians of science have sought to illuminate why certain categories are plausible and certain lines of inquiry are pursued, they have often identified underlying, perhaps implicit, patterns of thought and metaphors shared among different sciences and social thought more generally (Stepan 1986).[11] Gregg Mitman (1992) pursues this interpretive theme in *The State of Nature*—to choose an example relevant to this book's subject. Mitman describes the interactionist paradigm dominant at the University of Chicago during the first half of the twentieth century in the fields of animal physiology, embryology, behavior, and ecology: "Through its behavior and activity, the [biological] organism continually restructured its environment to meet new demands and thus had some control over the future course and direction of its development and progress" (Mitman 1992, 46). The central character in the book, W. Clyde Allee (1885–1955), clearly worked within this framework. Allee was an ecologist who received his doctoral training at Chicago before World War I and returned to join the faculty in the early 1920s. In his studies of animal aggregations, he sought to demonstrate that, across the many animal phyla and under a variety of conditions, individuals in groups withstood physical hardship longer than isolated individuals.

Chicago biologists applied the interactionist paradigm not only to their particular research, but to human affairs as well—and this is equally important to Mitman's account. In this spirit, Allee believed that "he had found experimental evidence opposing the doctrine of war, and also, the cornerstone to a theory of sociality centered not on the family but on the association of individuals for cooperative purposes found in the most primitive forms of life" (Mitman 1992, 4). Not only were Allee's scientific and social views connected, they correlated with his personal life and experiences as well. Allee was a Quaker and a pacifist who spoke out against World War I even after the United States joined the hostilities. His opposition to authorities and conventional hierarchies was evident in other ways, such as having a marriage that was a two-career partnership and attracting more female students to his lab than other faculty members did. In short, as Mitman narrates compellingly, Allee's science, biography, and social concerns were strongly correlated.[12]

In any field of inquiry, correlations raise the challenge of identifying causal links or mechanisms. In historical interpretation, one possible correlation-generating mechanism is suggested in the description of shared patterns of thought as *underlying* metaphors. The adjective implies that a pattern of thought widely shared in society has become *internalized* in the subjectivity of the agents in question and from this seat is expressed in all that those agents do—including, for scientists, in their scientific thinking. Personal life experiences obviously also shape an agent's subjectivity, perhaps influencing which patterns of social thought become internalized. A variant of this mechanism follows from seeing a metaphor not as something carried inside an agent's head, but as an aspect of language that people know how to relate to. Agents use metaphors in what they say and write because that makes it easier to communicate and influence others.

The broad line of explanation that focuses on metaphors and language does not, however, address the specific character of scientific practice. Allee and the other Chicago scientists performed experiments on real organisms, made observations, and analyzed results. How could these scientists do this research in ways such that historians (and sometimes the scientists themselves) could interpret the outcomes as supportive of their concerns about social order and their personal life histories? Several possibilities can be teased out:

1. When Allee spoke out about his social concerns, he appealed to selected parts of his science that he had done for conventional reasons. Similarly, he invoked parts of his science to justify the private and less public aspects of his life.

2. Allee chose particular scientific *questions*; namely, those whose results held promise of allowing him to make the appeals in 1.

3. Allee's life experiences and social concerns led him to shape his scientific *answers*, making possible the appeals to science mentioned in 1.

4. Allee's social concerns evolved so as to match the scientific results he was obtaining, making it easier for him subsequently to do 1, 2, and 3.

5. Among the many possible scientists Mitman could have written about, he chose special cases; namely, those such as Allee who happened to do 1, 2, 3, or 4.

What can we make of these possibilities? The last appears not to be the true—many other biologists during the period Mitman discusses sought to socialize nature and naturalize society. In any case, even if Mitman's choice of Allee were not representative of American biologists, we could still want to understand the causal relations producing the *social-personal-scientific* correlations for Allee. Of the remaining possibilities—all evident in Mitman's account—1, 2, and 4 are not hard to marry with conventional images of scientists who happen to have overt social concerns: These scientists participate in the dialogue between theories and nature as well as a dialogue with other scientists to establish knowledge, but the first dialogue can be separated from the second. However, possibility 3—that Allee's scientific *answers* were shaped by his life experiences and social concerns—challenges that separation. Through what practices of experimentation, observation, modeling, and so on did Allee bring about this shaping?

The problem taken up in this chapter is to identify a correlation-producing mechanism for possibility 3 that incorporates the specifics of a scientist's practice. I explore this problem not by revisiting Allee's work, but by picking up a strand of the study of ecological complexity that developed mostly after World War II; namely, systems ecology. My narrative begins with a vignette from the Great Depression, but centers on social and technical changes during and after the war. Against this background I identify and make sense of the social-personal-scientific correlations in Howard T. Odum's pioneering work in systems ecology during the 1950s and 1960s. This case allows me to propose a means through which the sociality of science can have a systematic effect on its referentiality.

Technocratic Optimism

In 1933, Howard Scott, founder of the Technocracy movement, offered his audience the choice between "Science or Chaos":

> I have not inquired as to whether you do or do not like the idea [of Technocracy]. The events that are going to occur . . . within the very near future are not going to be respecters of human likes or dislikes. The problem of operating any existing complex of industrial equipment is not and cannot be solved by a democratic social organization. . . . [It] is a technical problem so far transcending any other technical problem man has yet solved that many individuals would probably never understand why most of the details must be one way and not another; yet the services of everyone . . . will be needed.

The Technocracy movement commanded immense popularity for a brief period around the transition from the Hoover administration to that of Franklin Roosevelt (Akin 1977). It overshadowed all other proposals for dealing with the crisis of the Great Depression. The Technocrats proposed to replace what they called the "price system," which they saw as complex, unstable, and arbitrary, with equal allocations of nonaccumulable energy certificates. All materials and work could be measured in energy units; engineers, capable of making measurements free from the distorting interests of economics and politics, would organize society better than politicians. In fact, the transition to a Technocracy would also occur, the Technocrats hoped, without politics—once it was understood that their proposals would benefit society, the force of logic would ensure that those proposals would be implemented. A Technocracy, like previous proposals for industrially based utopias, would require individual discipline and participation; in return, order would be restored to a world perceived as rapidly changing and disordered (Segal 1985). The popularity of their movement was transient, but the Technocrats' vision of reducing society to a single energy dial, to be adjusted objectively by social engineers, would reoccur in the field of ecology.

In October 1946, the Yale ecologist G. Evelyn Hutchinson (1903–1991) delivered a paper entitled "Circular Causal Systems in Ecology" to an interdisciplinary conference at the prestigious New York Academy of Sciences (Hutchinson 1948). Hutchinson emphasized themes that would come to dominate ecology in the United States. In brief, he was exploring, as his title indicated, the concept of ecological relations as systems. This concept drew on, but also made significant extensions to, the then-prevailing organicist accounts of ecological complexity. Hutchinson's paper provides a convenient starting point from which I can trace conceptual connections and changes after World War II in the analysis of ecological complexity. I subsequently move my focus to a student of Hutchinson, H. T. Odum (1924–2002), who extensively developed one side of Hutchinson's program during the 1950s and pioneered the field that came to be known as systems ecology.

Organicism, undergoing a partial transformation into a systems view, was at the same time a source of social thought; ecological and social concepts were strongly connected in Hutchinson's and Odum's thinking. Their work allows me to highlight aspects of their social context—in particular, the *technocratic optimism* of the postwar years. The idea of technocratic management of society had a long history, but World War II, particularly as it was experienced by scientists, transformed the character of that political fantasy. Government funding and organization of science under military imperatives produced significant results, giving currency to the belief that intervention on a large scale could be practically realized. Moreover, scientific control of complex systems seemed necessary to prevent further social upheavals or holocaust. Optimism about the benefits of such control overshadowed possible doubts about its implications for democratic political life.

The term *technocrat* has come to denote not only followers of Howard Scott, but anyone advocating technical measures to address social problems. Technocrats believe they can handle social complexity in a value-free manner, maintaining a distance from specific interests and political details.

Through such nondependency and disengagement they can best serve all. But, as is typical of social philosophies framed in terms of universal interests, proponents build a special place for themselves in the proposed social organization. In interpreting H. T. Odum's work on ecological complexity, I show that technocratic optimism facilitated his interventions—actual or envisioned—in nature. Technocratic optimism was more than the context within which his work emerged; it facilitated his concepts, methods, and organization of research, as well as the ways in which these aspects of his work facilitated one another. Odum was enabled not only to *think* that ecosystems were *like* well-designed feedback systems, but to *act,* in many ways, *as if* they were.

Circular Causal Systems in Ecology

The paper on "Circular Causal Systems" by Hutchinson, who would become Odum's mentor, had two sections, disjunct in content and tone. The first section described "the biogeochemical approach," an intersection between biology and the study of the global distribution of chemicals. This approach to the study of ecology was mostly dry and quantitative: Hutchinson constructed budgets of carbon in the biosphere (fig. 3.1) and of phosphorus in lakes, and attempted to balance those budgets. Biological and physical processes were tightly linked: he discussed the activity of organisms in terms of their balancing effect on cycles of chemicals through organisms, the earth, oceans, and the atmosphere, and he related changes in biological productivity to changing concentrations of available nutrients.

The second section, on "the biodemographic approach," was speculative. Hutchinson developed this approach through mathematical equations proposed for population growth with little reference to observations. The modeled populations met face-to-face, or variable-to-variable, and their mode of interaction and regulation was reduced to single mathematical parameters denoting reproduction, competition, or predation. The use of simple models led to the paper's culmination: a graph of scientific knowledge as a variable undergoing faster-than-exponential growth.

The two sections of the paper differed markedly, yet in the brief introduction Hutchinson provided a basis for their unity, as well as for the unity of divergent currents within ecology. The "conditions under which groups of organisms exist," he remarked, should be seen as systems of circular causal paths, or feedback loops. These circular causal paths were "self-correcting within limits" (Hutchinson 1948, 221). When the limits were exceeded, violent oscillations would drive some elements of the system to

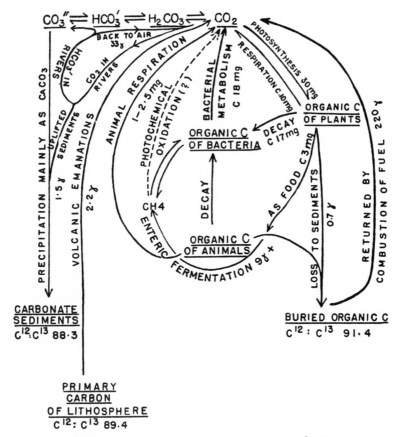

Figure 3.1 The global biogeochemical cycle of carbon. Quantities are per cm² of the earth's surface per year. Arrows represent general pathways of the flow of carbon. Paired arrows represent reversible reactions; dashed arrows represent less well-established pathways. (From Hutchinson 1948, 223.)

extinction. The original system would thus be replaced by a new one in which the lost element would play no part. Hutchinson drew the inference that natural systems would, therefore, have mechanisms to damp oscillations and thereby ensure their persistence. This self-regulation might be purely physical—for example, excess atmospheric carbon dioxide dissolving in the ocean—or it might be primarily biological—for example, organisms in limited space having an inhibitory effect on one another, as expressed in the terms of the logistic equation for population growth.

For Hutchinson, whether ecology was biogeochemical or biodemographic—or, in other words, whether ecology was descriptive and quantitative or abstract and mathematical—it was nevertheless united by a theoretical proposition: Groups of organisms are systems having feedback loops that ensure self-regulation and persistence.

Conceptual Connections: G. E. Hutchinson

Hutchinson's paper presents an almost perfect study for a history of scientific ideas. His conceptual precursors can easily be traced: the first biogeochemists, V. Goldschmidt and V. Vernadsky (see Vernadsky 1944, which was translated and condensed by Hutchinson), and the biodemographers V. Volterra, A. Lotka, and F. Gause (Kingsland 1995). The contemporary connections are also clear, both within ecology and through a series of influential meetings sponsored by the Macy Foundation. And the paper also foreshadows the future, almost uncannily. Hutchinson's biogeochemical approach, developed by his student H. T. Odum, becomes systems ecology. The biodemographic approach becomes community ecology, with another of Hutchinson's students, Robert MacArthur, becoming responsible for the direction this field took after the late 1950s (chapter 1; see also Kingsland 1995, 175–205). With an eye toward H. T. Odum's work, I will leave aside Hutchinson's precursors as well as MacArthur and community ecology to concentrate on the biogeochemical approach to the study of ecological systems.

Hutchinson stressed the importance of analysis in ecological studies. His quantitative, chemical approach contributed to the analysis of ecological systems. Hutchinson criticized ecological principles that maintained a dualism between living formations and their physical conditions—climate, soil and so on—and he worked to integrate biological and physical processes in a practical program of ecology. Hutchinson placed his analytic approach in contrast to the increasingly elaborate schemes of his contemporaries: schemes for classification of ecological communities, of their sequences of development or succession, and of the end points or climaxes of those successional sequences. Similarly, he wanted to go beyond discussion about whether communities were organisms at some higher level. When he reviewed *Bio-ecology,* a textbook of general ecology published in 1940 and written by the influential ecologists Clements and Shelford, Hutchinson had commented that their "general principles are . . . mainly classificatory." He concluded that "if . . . the community is an organism, it should be possible to study its metabolism" (Hutchinson 1940).

In the spirit of ecologist analyzing ecological metabolism, Hutchinson had enthusiastically sponsored Raymond Lindeman's tragically short career in ecology and encouraged his analyses of energy flow through trophic levels of ecological communities; that is, through plants, herbivores, and carnivores (Cook 1977). In a commentary to Lindeman's classic (posthumous) 1942 paper, "The Trophic-Dynamic Aspect of Ecology," Hutchinson remarked that Lindeman "came to realize . . . that the most profitable method of analysis lay in the reduction of all the interrelated biological events to energetic terms" (Hutchinson 1942, 417).

Hutchinson's emphasis on analysis meant more, however, than measurement of energy and elemental flows; he had begun to search for a theoretical basis for that analysis. He commented that the formulation of Lindeman's energy analysis provided "a hint of some undiscovered type of mathematical treatment of biological communities" (Hutchinson 1942, 418). Here was an early sign of Hutchinson's biodemographic approach; in fact, his interest in mathematical ecological theory had already led him to draw on the work of Gause, Volterra, and Lotka (Kingsland 1995, 179).

Hutchinson's search for a theoretical basis for ecology was coupled with catholic interests, writing, and acquaintances (as evidenced in "Marginalia," a regular feature he wrote for *American Scientist* from 1943 to 1957). This had led to his participation in the Macy conferences, a series of interdisciplinary meetings from 1946 to 1953 sponsored by the Josiah Macy Jr. Foundation. "Circular Causal and Feedback Mechanisms in Biological and Social Systems" was the original name of the meetings; later this was shortened to "Cybernetics" (Heims 1991). The 1946 conference on "Teleological Mechanisms" at the New York Academy of Sciences was, in effect, a Macy conference "open house." The introduction was given by Lawrence Frank, a social scientist and an instrumental figure in the Macy conferences and in other interdisciplinary ventures. Frank was followed by Norbert Wiener, the "father" of cybernetics; then Hutchinson spoke. Warren McCulloch, initiator and chairman of the Macy meetings, concluded the conference. All of these figures will contribute to my account in due course.

The Macy conferences had an enormous influence in many fields, popularizing the perspective that complex systems can be treated as self-regulating feedback systems. The impact extended well beyond the diverse and unresolved issues discussed during the meetings (Heims 1991). The conferences catalyzed the transfer to peacetime of an optimism born of the experience of wartime science. During the war, scientists, including many of the Macy participants, had worked under intense demands, but in a new environment of greatly increased resources, cooperation among scientists, and social recognition.

Frank expressed this optimism at the New York Academy meeting: "[W]e are engaged, today, in one of the major transitions or upheavals in

the history of ideas. . . . When the social sciences accept these newer con-
ceptions . . . and learn to think in terms of circular causal processes, they
will probably make amazing advances" (Frank 1948, 192, 195). Students of
the many fields represented at the Macy conferences believed that it
would be possible to construct a theory about complex systems. Moreover, such a theory might unify the physical, biological, and social sciences, allowing the success of physics to flow into other fields (Bateson
1946; Hutchinson 1946). It was in this context of optimism about a scientific approach to society that Hutchinson was developing his views about
ecology and systems.

Organisms and Systems

Hutchinson viewed systems thinking as a development in the theory of the
complexity of ecological relations. He was critical, as I noted, of the organicist emphasis in ecology, wherein a community of different species was
held to correspond to some complex organism that, like an individual
organism, had unity, had interdependent parts, and developed over time.
His criticisms notwithstanding, Hutchinson and his community-oriented
contemporaries shared a central precept, that nature was divisible into integrated wholes. Regardless of whether the whole was a community or a biogeochemical cycle, there were conceptual steps made in common, steps
corresponding to both methodology and worldview. These ecologists had
to propose boundaries to delineate a unity or whole and articulate its components; for example, the plant populations making up a vegetation formation or the stores of carbon in the global carbon cycle. The internal
relations—that is, the relations they postulated among the components—
were considered coherent and distinct from the external factors determining the behavior of the community or system. In this sense, Hutchinson's
self-regulating systems constituted a small step from the homeostatic communities of his ecological contemporaries; circular causal paths were likewise a small step from groups of interspecific populations acting, reacting,
and coacting to ensure coordination within the communities (Clements and
Shelford 1939).

   In several ways, however, feedback systems marked a departure from
the prevailing idea of communities as organisms. In the systems view, living and nonliving feedback systems alike obeyed common mechanical
principles. These principles included their mode of evolution and self-
regulation. Data could be directly used to elucidate the dynamics of systems. And, once scientists understood the dynamics of systems, those

systems would be controllable, enabling society to become free from catastrophe. I will draw on the papers of Hutchinson and the others at the New York Academy conference to explore these four aspects of the change to a systems understanding.

Unifying the study of living and nonliving systems was still a radical step, even though vitalism was a defeated force in biology. The new theorists of feedback systems conceived of nature as a machine and, at the same time, acknowledged the purposive and regulatory character of that nature-machine. A theory of so-called teleological mechanisms could abolish not only vitalism, but also the old cause-effect determinism (Frank 1948). Furthermore, the same terms could be applied to all systems, whatever their components; living and nonliving could be intermeshed, eliminating the separateness of biological relations from physical factors (Wiener 1948b; Hutchinson 1949). Hutchinson's ecology emphasized chemical and physical processes as much as biological, the geochemical as well as the demographic. It would not be long, as I shall show, before purely physical theories, such as those of thermodynamics, or even more abstractly, of information theory (Margalef 1958; Patten 1959), would be taken up as organizing principles for ecology.

The unification of living and nonliving systems resulted in a changed picture of evolution and self-regulation. In organicist ecology, communities-as-organisms evolved by natural selection. Selection's firmly guiding hand improved the adaptation of different species to one another. The coordination of species was necessary to ensure the existence of the community as an entity; in other words, species served a function in the maintenance of the community organism. Alfred Emerson, writing in *Principles of Animal Ecology*—the synthesis of organicist ecology—provided the mature expression of this view: "ecological evolution parallel(s) the evolution of internal physiological balance and control within the organism" (Allee et al. 1949, 598; see also Mitman 1992). Feedback systems, on the other hand, achieved self-regulation and stability by a process applicable to all systems, living and nonliving. In contrast with Emerson's superorganismic view, Hutchinson's analysis of ecological metabolism brought actual ecological compartments and flows into focus. Admittedly, the persistence of components of ecosystems was still dependent on the self-regulation of the whole system. In Hutchinson's view, oscillations of systems that failed to self-regulate drove to extinction those components that were destabilizing (Hutchinson 1948, 221). While this formulation might seem like selection in another form, Hutchinson did not refer to it as such. It was not the intimately superintending natural selection invoked by organicists; the

individual-in-system gained some autonomy relative to an individual fulfill-
ing its function in a community. This was a shift in emphasis, not a conceptual break. After all, Hutchinson's concept of system stability was no less
powerful a criterion for persistence than was Emerson's idea of group
homeostasis. In fact, given its applicability to all systems, it was potentially
more effective.

A third feature in the transformation to feedback systems was the power
of data, or, at least, the promise of its power. In the place of descriptive and
classificatory approaches to complexity, scientists could use long-time runs
of data to detect feedback and expose the dynamics of systems. At least,
that was the implication drawn from Wiener's (1948a) theories about time
series.

> All we have a right to ask of the appropriate sciences are long-time runs of data.
> We know it will take years to collect these, but we must have them before we can
> determine whether the mechanism of negative feedback accounts for the stability
> and purposive aspects of the behavior of groups. (McCulloch 1948, 264)

With these concluding remarks to the New York Academy meeting,
McCulloch foreshadowed a new social science based on analysis of data, a
science that has subsequently flourished with the advent of high-speed
computers. McCulloch's vision, self-confessedly utopian, was "that man
should learn to construct for the whole world a society with sufficient inverse feedback to prevent another and perhaps last holocaust"
(McCulloch 1948, 264). On a more prosaic level, Hutchinson expressed a
similar view in remarking that Lindeman's measures of energy content in
the different levels of ecosystems and in the flows through those levels provided a basis for the abstract analysis of the dynamics of ecosystems
(Hutchinson 1942, 418). In other words, the descriptive measures could be
translated into dynamic models. This translation from data to dynamics
would become important in systems ecology.

A final aspect of the transformation from community organisms to
feedback systems was illustrated in the vision of a cybernetic social science heralded by McCulloch. Freedom from holocaust, and from other
social upheavals, might be achieved through the construction of an all-
encompassing system of feedback. A systems approach to understanding
nature moved easily into a systems approach for engineering society. But
who would be the engineers of social systems? This question, not explicitly
asked by the Macy participants, would find one answer in the developing
work of Hutchinson's student, H. T. Odum.

Conceptual Connections: H. T. Odum

Howard Thomas Odum and his older brother Eugene (1913–2002), who became an ecologist before him, were the sons of Howard Washington Odum (1884–1954). Their father, an influential sociologist, held profoundly organicist views and proclaimed himself the standard-bearer of Lester Ward's dynamic sociology. (The endpapers of his 1947 text display the lineage from Ward to Odum.) H. W. Odum was best known for his work on Southern regionalism and the folkways of Southern blacks, and for his efforts to promote the cooperation of intellectuals and other social groups in the rebuilding of the South and its reintegration into the nation (Brazil 1975). Of the two sons, Eugene Odum was the better known among ecologists. His textbook *Fundamentals of Ecology* (Odum 1953, 1959, 1971) and his founding of the Institute of Ecology at the University of Georgia kept him in a central position in the academic discipline of ecology (Hagen 1992; Kwa 1989). Nevertheless, the conceptual and practical developments that interest me, many of which are evident in Eugene Odum's writing, originated with the younger brother.

H. W. Odum encouraged his sons to go into science and to develop new techniques to contribute to social progress (Odum 1986; see also Brazil 1975, 386). H. T. learned his early scientific lessons from several sources: his brother taught him about birds; the marine zoologist Robert Coker, for whom he worked after school, taught him about fish and philosophy of biology (Coker 1939); and *The Boy Electrician* taught him about electrical circuits (Odum 1986). His college education was broken by three years of Air Force service as a tropical meteorologist in Puerto Rico and the Panama Canal Zone, after which he returned to complete his studies in zoology and chemistry.

After the war, Hutchinson was looking for graduate students with a background in physics and chemistry. In 1947 he took on H. T. Odum as a student and steered him to the study of the biogeochemistry of strontium (Odum 1950). Hutchinson's suggestion of this topic derived from his systematic study, following Goldschmidt, of the biogeochemistry of different elements; neither Hutchinson nor Odum anticipated the interest in strontium generated by later atmospheric testing of atomic weapons. Odum's dissertation, completed in 1950, indicated that the chemical composition of the oceans had not changed in the last 40 million years, at least insofar as it would alter the ratio of calcium to strontium—a finding in contrast to the prevailing wisdom. This result was labeled one of the top twenty scientific discoveries of the year in a *Life* magazine feature on U.S. science (*Life* 1950).

Odum's dissertation exemplified Hutchinson's biogeochemical approach and concept of systems as expressed in the "Circular Causal Systems" paper. Odum's approach was interdisciplinary; "any parameter that cuts through science boundaries is indeed welcome," he intoned (Odum 1950, 333). Observations and data were drawn from a wide range of fields in an attempt to "connect isolated facts" into "one coherent and quantitative system" (Odum 1950, 1). Data about a system and that system's dynamics were closely connected: constancy of the pattern of distribution of strontium was interpreted as evidence of the self-balancing dynamics of the strontium ecosystem—an entity of "worldwide dimension" (referring to Vernadsky's concept).

In formulating principles about systems, Odum, like Hutchinson, made no distinctions between living and nonliving processes. For example, the "stability principle," which Odum attributed to Lotka, ensured that "nature is as a whole in a steady state or is in the most stable form possible and constitutes one big entity" (Odum 1950, 9). In this view, natural selection in inorganic systems and at higher levels of organization did not require differential survival in competition among variant entities, such as ecosystems. It was sufficient that "a system which has stability with time will exist longer than a system without stability" (Odum 1950, 8).

Odum attended one of the Macy meetings as a guest of Hutchinson to talk about his strontium research. The discussion, as he later recalled it, was unruly and unilluminating (Odum 1986). Nevertheless, his affinity with the Macy conferees' new, *teleological* view of mechanism was clear. In his dissertation he described ecology as one part of the study of "mechanisms of steady states in all types of system," quoting Wiener's definition of cybernetics. Odum adapted the marine ecologist George Clarke's (1946) picture of food chains as a set of cogwheels, expanding this concept to a biosphere of cogwheels, "each constituent chemical cycle a cogwheel that is geared to other cogwheels, the whole system being interconnected" (Odum 1950, 308).

My tracing of conceptual connections would be incomplete without mention of Odum's many references to Lotka's work, in particular *Elements of Physical Biology* (Lotka 1925). Many ecologists, including Hutchinson, have referred only to Lotka's mathematical models—indeed, "Physical" was replaced in the title by "Mathematical" in later reprints of the book (Lotka 1956). Odum, however, grasped the intent of Lotka's original title; namely, the analogy of physical biology with physical chemistry. In one of Odum's few references to organisms, he called them "ecocatalysts," able to "lower the free energy of activation" of each step in a cycle so that the system would reach a different equilibrium than it would without them (Odum 1950, 325). Odum's intellectual debt to Lotka will appear even stronger in my later discussion.

Two aspects of the systems thinking of Hutchinson and the Macy conferences were, it might be noted, missing from Odum's dissertation. The first was the inclusion of humans in the biogeochemical systems and their possible reordering of those systems. The second was Hutchinson's enthusiasm for energetic analysis of ecosystems. These themes, however, would emerge during the next decade of Odum's career.

Ecosystems as Energy Circuits

In 1951, Odum published a short article based on his dissertation, which emphasized that stability was the fundamental property of the strontium cycle (Odum 1951). The stability of the cycle and its inclusion of both living and nonliving components led Odum to call the strontium cycle an *ecosystem*. Significantly, a minor comment from his dissertation "emphasiz[ing] that these cycles are not energetically closed" (Odum 1950, 311) was elevated to a prominent introductory position in this article: biogeochemical cycles are "driven by radiant energy" (Odum, 1951, 407). From that point on, energy drove Odum's development of the biogeochemical approach to ecology.

Odum's first research project as a young professor at the University of Florida at Gainesville was a study of the trophic structure and productivity of the Silver Springs in central Florida. This project continued Lindeman's research program to measure efficiencies of energy transformation from plants to herbivores to carnivores to higher carnivores. In addition, Odum advanced several suggestive hypotheses about biological communities and the partitioning of energy. For example, he proposed that the ratio of total community production to total community respiration might determine the character of the biological community (Odum 1955).

Odum placed this research within the systems perspective (see H. T. Odum in E. P. Odum 1953, 67), stating that he aimed to identify the control mechanisms by which the states and flow of energy are sustained (Odum 1955, 59). Nevertheless, Odum, like Hutchinson, retained many terms and emphases of organicist ecology. For example, in another passage he said that his goal was the study of "mechanisms by which the community metabolism is self-regulated" (Odum 1955, 56). Odum's Silver Springs research and, I suspect, his organicist perspectives were encouraged by W. C. Allee (Mitman 1992), who had just moved to Florida from Chicago to head the zoology department at Gainesville (Odum 1986). Some of Odum's organicist perspectives would later lead many systems ecologists to distinguish their work from that of Odum and his brother, but not before H. T. Odum had made several innovations on which systems ecology is based.

During his Silver Springs project, Odum began to draw energy flow diagrams of ecosystems (fig. 3.2)—an obvious step, so he claims, for someone with biogeochemical training (Odum 1986). (The first flow diagram appeared in Odum 1956b.) With Odum's background in electrical circuitry, it was a small step to convert these energy flow diagrams into actual electrical circuits, an analogy Odum would explore until the analogy dissolved into equivalence (Odum 1962b). Variable resistors could be adjusted until the current flow in an electrical analog circuit (fig. 3.3) became proportional to the measurements of energy (or its biomass equivalent) flowing between trophic compartments in the actual ecosystem. The voltage drop between compartments suggested to Odum the analogous "ecoforce." Among the hypotheses stimulated by his manipulation of electrical circuits was his suggestion that ecologists think about available "food by its concentration practically forc[ing] food through the consumers" (Odum 1960, 5).

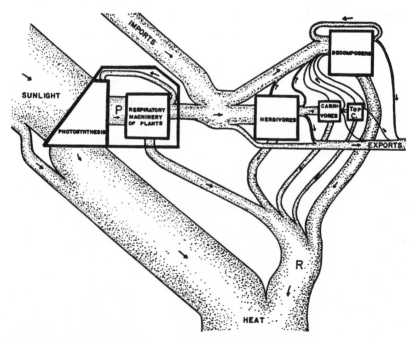

Figure 3.2 Energy flow diagram for an ecosystem. (From Odum 1956b.)

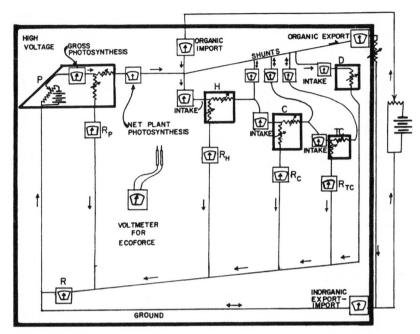

Figure 3.3 Electrical analog circuit for the steady state ecosystem in Figure 3.2. The zigzag lines with arrows represent variable resistors. (From Odum 1960.)

The use of electrical circuits in turn inspired Odum to develop a set of symbols that he hoped would provide a common language ecologists could use to discuss energy flow and electrical systems theorists could use to convey their synthetic insights to ecologists. The symbols, refined during the 1960s, would become his hallmark in subsequent publications (Odum 1982).

Odum's energy circuit diagrams were readily generalized. During the 1950s he had taken the necessary measurements from diverse situations: a coral reef at Eniwetok Atoll in the Pacific (Odum and Odum 1955), microcosms and ponds in North Carolina, estuaries in Texas (Odum and Hoskin 1958), and a tropical rain forest in Puerto Rico. When he used energy diagrams to summarize the functioning of these different ecosystems, Odum hoped they would indicate where there was similar function despite the taxonomic differences between ecosystems or locations within an ecosystem. He emphasized general principles instead of the particularities of the organisms and their interactions—a "functional ecology," as popularized in his brother's textbook. For H. T. Odum, measurement in a common currency—energy—might make it possible to discover universal principles of ecosystem "design," including not only the energy transformations of living matter, but also those of physical processes such as erosion, which moves the sedimentary cycle, or wind, which influences evapotranspiration from leaves of trees.

Measurement was central to Odum's ecology, as it was to systems ecologists who followed him. By collecting data for an entire system and summarizing them in flow diagrams, the systems ecologist could act as if the diagrams represented the system's dynamic relations. This approach, it should be noted, might require some arbitrary internal aggregations, such as summing species into trophic compartments. The flow diagrams, when transformed into computer models, could be used by systems ecologists to generate predictions about the future or about responses to perturbations. These kinds of data have retained a powerful hold on the imaginations of Odum and other systems ecologists. Redescription, or bookkeeping, of measurements on an ecosystem has frequently been used as if it provided a representation of the ecosystem's dynamics; that is, of the ecological relations that generated the observed data.

While other systems ecologists would come to measure variously biomass, population sizes, energy, or essential elements such as nitrogen, Odum converted everything into energy. This currency had a special status. All organisms require energy, and, following Lotka's lead, Odum expected theoretical generalizations in ecology to take the form of biological additions to the thermodynamic principles of physical chemistry (see H. T. Odum in E. P. Odum 1953, 67). By 1955, Odum had formulated the "maximum power principle" (Odum and Pinkerton 1955), which he would later describe as the "fourth law of thermodynamics" applicable to open systems (Odum 1960, 1). (In a closed system, the second law of thermodynamics dictates that entropy—dissipated or unusable energy—increases over time.) The maximum power principle was the complement of the energy circuit–electrical circuit analogy; together, they formed the physical basis of Odum's ecology.

The maximum power principle combined the stability principle mentioned earlier, which Odum attributed to Lotka, with another theoretical suggestion Lotka made about energy and natural selection. Lotka had proposed that evolution "proceeds in such direction as to make the total energy flux through the system a maximum compatible with the constraints." This principle could be "derived upon a deductive basis" from the principle that, all other things being equal, "in the struggle for existence, the advantage must go to those organisms whose energy-capturing devices are most efficient" (Lotka 1922, 147, 149). Competition among organisms disappeared, however, in Odum's version. Instead, the maximum power principle was stated in terms of persistence—that is, as a variant of the stability principle: survival of the fittest meant "persistence of those forms which can command the greatest useful energy per unit time (power output)" (Odum and Pinkerton 1955, 332).

Odum's ecological theory relied on the universality of the maximum power principle. In his original paper on that principle, written with chemist Richard Pinkerton, he applied the same general formulation to many cases: a waterwheel driving a grindstone, one battery charging another, food captured by an organism for its maintenance, primary production in a self-sustaining climax ecological community, and the growth and maintenance of a civilization (Odum and Pinkerton 1955, 337ff). The maximum power principle covered any open system exhibiting self-reproduction and maintenance, irrespective of its scale, the placement of its boundaries, or its internal aggregation into compartments. Models from population, community, and systems ecology alike could be reexpressed in energy diagrams; the maximum power principle implied that this was the natural way to develop such a synthetic and comparative reduction of otherwise bewildering complexity.

Since Odum considered the maximum power principle to be universal, he urged ecologists to pay special attention to systems that, through isolation and lack of disturbance, have adjusted so as to achieve optimum power. From rain forest and tropical reef ecosystems (Odum and Odum 1955), for example, humankind could learn "about optima for utilizing sunlight and raw materials." Moreover, Odum recognized that human intervention in previously undisturbed ecosystems, whether desirable or not, was taking place: "the old systems and the new are being joined into an overall network including factories and towns, reefs and grass flats, and the flows between them" (Odum 1967, 99–100). Nature could teach us how to design well these "systems of man and nature" (Odum 1967; see also Odum 1962a). In this spirit, Odum initiated in 1957, with funds from the Rockefeller Foundation, an ambitious project intended to show how a mature natural system—a tropical rain forest in Puerto Rico—evolves or modifies its own design in response to a massive input of energy. From this study he hoped to derive a model for humans to follow.

The goal of designing "systems of man and nature" was leading Odum outside the boundaries of the academic discipline of ecology. The Puerto Rican rain forest project, nevertheless, became a practical model for subsequent systems ecology. In 1962, the Atomic Energy Commission, desiring a study of the "consequences of nuclear warfare or major reactor accidents"

(Bugher 1970), committed $360,000 to the project. Under Odum's scientific direction, a wide variety of scientists collaborated—albeit somewhat loosely—on the study of one ecosystem. The project would create a picture of the forest before and after exposure to three months of irradiation—in Odum's terms, an input of additional energy. Odum had pioneered the highly managed, collaborative, integrated ecosystem research project that commanded significant government support. Ecology had entered the domain of "big science" (Weinberg 1961).

After the rain forest study, Odum continued to be a successful and productive scientist—recognized, though not always understood in the way he would like, well beyond the boundaries of the academy (see, for example, *Newsweek* 1975). By the time he retired—though that is hardly an apt term given his continued activity—he had written 10 books and over 300 articles, and he directed, after he returned to Florida in 1970, over 100 master's and Ph.D. thesis projects. His research, including further large-scale interventions in ecosystems, continued to attract collaborators from around the world and funds from a variety of sources (see Hall 1995, especially fig. P.1).

In many respects, the new discipline of systems ecology followed Odum's leads. For example, during the International Biological Program, a massive ecological research endeavor that began in the late 1960s (Worthington 1975; Kwa 1989; Hagen 1992; Bocking 1997), ecologists undertook large, integrated research projects to accumulate measurements on entire systems, appropriately bounded and divided into compartments. Storages of biomass or specific nutrients in the compartments and flows between those compartments were translated into computer models. That is, data were directly converted to dynamics. Ecological complexity seemed to be reducible in a systematic manner; systems ecologists could move from system to system, collaborating in making compartment-level measurements that reduced the need for intimate knowledge of the range and flexibility of particular species' behavior. In these features, systems ecology and Odum shared a similar technocratic orientation.

In other ways, however, the mainstream of systems ecology diverged from Odum's approach. Most systems ecologists were skeptical about systems evolving to achieve optimum function—that is, maximum power. Odum's extensive use of the stability principle seemed unnecessary; his focus on energy overly restrictive. General theoretical principles were of less importance than an overall methodology for organizing and completing projects on a huge variety of systems. (By the 1980s, the lack of general theory haunted systems ecology. A school of hierarchy theorists has attempted to remedy this shortcoming: Allen, O'Neill, and Hoekstra 1984; O'Neill et al. 1986.) At times, proponents claimed that systems ecology would lead to sound environmental management. Nevertheless, in practice, it focused on basic science (National Academy of Sciences 1974, 5; Bocking 1997); the more ambitious project of developing a theoretical basis for human intervention in nature was postponed (Kingsland 2005). As we shall see, such a pragmatic separation was untenable for H. T. Odum.

From Organicism to Technocratic Optimism

Designing "systems of man and nature" resonates with the vision of a new social science that McCulloch had announced at the New York Academy meeting: freedom from holocaust and other social upheavals might be achieved through the construction of an all-encompassing feedback system. The possibility of controlling complex systems marked a simultaneous development in ideas about social organization and in representations of nature. Nature and society shared a universal feedback logic, a logic consciously and dramatically materialized in the wartime technology of anti-aircraft missile guidance. Odum, like the Macy conference participants, sought to naturalize society and socialize nature.

Recall the problem posed at the start of this chapter: By what mechanism do scientists' life experiences and social concerns get into their scientific theories? Let me now develop my answer by analyzing how the Macy conferees and social and natural theorists before Odum responded to social and technological developments that changed their social status, sometimes threatening, sometimes enhancing the modes of social action with which they identified most strongly. The transformation of natural and social theory from community organisms to feedback systems can be read as the theorists' attempt to make sense of their experience as social agents. Their actions as well as their ideas regarding nature, machine, and society animate one another (see note 11 on metaphors). Under this interpretation, it is difficult to demarcate a social realm *outside* science that has to get *into* the science (Haraway 1984–1985, 53).

In Howard Scott's words, technocracy alone offered life (Scott 1938). Technological development had made the Technocratic social order possible—vast increases in energy utilization allowed Technocracy to promise a short workweek for all. At the same time, technological change had made a new order necessary—industrial production had become so complex and interdependent that the failure of any one component could disrupt the entire "machine." The Great Depression and idle productive capacity proved to the Technocrats and their supporters that the organization of industry had broken down. Only a cadre of engineers using scientific principles could solve the technical problem of restarting and running the industrial machine at maximum efficiency.

In many respects, the Technocrats' polemic simply expressed an extreme current of the organicist mainstream. Interdependent components of the machine are like the organs of the social organism—both function in a natural division of labor. In the early 1930s, many social theorists viewed the machine as broken; the organism was sick or out of balance. The organicists' diagnosis of this crisis was, like the Technocrats', one of "social lag": the pace of scientific discovery and technological innovation had become so rapid that the necessary adjustment of social institutions and sentiments lagged behind (Ogburn 1922; Cross and Albury 1987).

Social balance had been the central problem for the organicists even before the crisis of the Great Depression. The Progressive period and World War I had advanced the cause of state regulation and planning, opening up the possibility of new administrative and managerial roles. Organicism was to a large extent the discourse through which leaders and administrators defined their roles in the post-Progressive period and responded to social and political forces that were not easily balanced. Immigration, for example, was viewed as a threat to the dominant status of these primarily Anglo-Saxon American men. Immigrants brought with them different cultures, religions, and, it was claimed, diseases as well as differential reproductive rates. They had to be assimilated into a "traditional" American way of life, or contained without provoking revolt. Class conflict had to be averted, and this meant restraining industrial capitalism or mitigating its excesses. With the crash of 1929, this issue became urgent.

The language of the organicists was not, of course, so explicit about their interests; their silences are instructive. Progress or social change was a force curiously external to the social organism. Capitalism and science as prime movers of social change were afforded, therefore, a privileged existence. Although social lag was viewed as a result of the rapid development of science and industry, it was society and its components that needed to adjust. Terms such as adjustment, adaptation, integration, and function dominated the discourse of the organicist or functionalist social sciences. Cooperation among the specialized parts of the social organism was viewed as essential to the survival of the whole, which, in turn, would ensure the welfare of the parts. (This was similarly the case when Emerson, W. C. Allee, and others advanced biological bases for ethics in response to World War II; Mitman 1992.)

Yet the cooperation sought by organicists could equally well have been described as subordination to a hierarchical division of labor. H. W. Odum—the ecologists' father—had discussed the "misunderstandings" that stand in the way of cooperation between laborer and capitalist (Odum 1927, 470ff; see also Haraway 1983). In the discourse of organicism, management had moved to the realm of necessity: natural spontaneity must be subjected to some social control; regulation of social processes was a precondition for free and individual life. Individuals who were well adjusted would recognize their responsibility to work in harmony and to ensure stability and order. Some degree of control was not only compatible with traditional democratic values, it was necessary for their defense (Cross and Albury 1987).

Those holding a common vision of a managed society disagreed about the degree of social control needed. The Technocrats, whose leaders were marginal figures politically, fantasized a radical redesign of society; they nominated engineers such as themselves to run the new industrial organization. The predominant organicist discourse, on the other hand, was sustained by leaders in the academy, industry, and government. It was more pragmatic, advancing an administrative or a therapeutic model that was informed, in the notable cases of L. J. Henderson and Walter Cannon, by work in physiology (Cross and Albury 1987). Change would not be revolutionary; it would be planned.

Organicist commentators implied a role for themselves as rational, practical administrators or, as H. W. Odum expressed it, leaders in "man's quest for social guidance" (Odum 1927). In Emerson's writings, the powerful administrator became naturalized in the role Emerson gave selection, guiding the improvement of the whole and ensuring group homeostasis (Mitman 1992). Nevertheless, organicists were not of one mind about the degree of intervention required; their differences paralleled the wider social debates about government intervention. The Hooverites emphasized private, voluntary planning and coordination. In this spirit, Henderson used case studies of concrete problems to train administrators, men who would become intuitively skilled in the redirection of human relations. Cannon, on the other hand, tentatively advocated a "biocracy," a noncentralized set of agencies that would regulate the processes of commerce (Cannon 1933). This picture foreshadowed Roosevelt's New Deal agencies, which increased, in measured steps, federal planning, regulation, and intervention to stimulate "national recovery."

Ironically, given the success of the New Deal in legitimizing rational management in place of "laissez-faire," there was no distinctive New Deal

science. Government support and long-range planning in science met successful resistance from many influential scientists (other than in agriculture,
for which there was a long tradition of government involvement). They
feared that science in a "planned economy" would lose its perceived independence from politics. Their insistence on the method of "private initiative" frustrated the physicist Karl Compton's energetic advocacy of a federal
science advisory agency with broad responsibility for funding science
(Kargon and Hodes 1985). World War II would change all that.

Late in 1944, President Roosevelt asked Vannevar Bush, a protégé of
Compton and director of the wartime Office of Scientific Research and
Development, for advice about government involvement in the extension of
wartime science to peacetime. Bush wanted scientific research after the war
to operate primarily in the framework of civilian institutions constituted to
be separate from the direction and secrecy of the military. In this goal
he was not successful; his report, *Science: The Endless Frontier* (Bush 1945),
was not the blueprint for postwar science policy that subsequent commentators have held it to be (Dennis 1997). Nevertheless, many of his proposals—
in particular, his distinction between basic and applied research—gained
currency in the debates that led eventually to the establishment of the
National Science Foundation. (Hutchinson [1945] applauded Bush's report
in a detailed review in *American Scientist*.) In short, Compton's frustrated
plan had been revived; some form of alliance between government and science was now taken for granted. Any resistance lingering from the thirties
had succumbed to the experience during the war of science that was generated by the federal government and served the national interest.

The war had not, however, swept away the prewar organicist themes of
social lag and adjustment. Lawrence Frank, the social scientist who would
become a leading figure in the Macy conferences, made this clear in his
report of a 1944 symposium on "Research after the War." Frank insisted that
a national policy of scientific research would, among other things,

> recognize that the very progress of research in physical science and technology
> made it imperative that research in the social sciences and the humanities, espe-
> cially into the traditional American patterns of thinking and action, be further
> developed and improved, since the growing discrepancy between our advancing
> technology and our established practices and organizations is one of the major
> threats to our free, democratic social order. The need for the exercise of critical
> thinking upon our folkways and our historically derived social, economic and
> political beliefs and patterns is no less than the need for critical thinking upon our
> industrial processes and technical equipment and practices. (Frank 1945)

Nevertheless, the preeminent role afforded to science, albeit enmeshed with the military, was transforming organicism. In particular, the scope of scientific intervention in society could be increased. Previously, when nature or society were viewed as an organism, the scientist's role was to describe and classify the natural order and the natural mechanisms of integration; perhaps the scientist could also advise on how the organism might be healed or might heal itself (Cross and Albury 1987). Systems, in contrast to organisms, could be constructed, or so McCulloch claimed; systems were constructs in which scientists could intervene and exert control.

A social feedback system, however, implied the existence of systems scientists under whose controlling hands the system would run for the benefit of the rest of society. The implication that control would be abdicated to a cadre of scientists was not a major concern to the Macy participants, even if some of them were skeptical on technical grounds about the possibility of social feedback systems (Wiener 1948a, 162). The new theories about feedback systems were liberating for the arms-race enthusiast John Von Neumann and, at the same time, for humanists such as Hutchinson, Margaret Mead, and Gregory Bateson (Heims 1980, 1991). In 1946, Bateson called on physicists and other natural scientists to help prevent atomic war, proposing a mission for the social sciences inspired by the Manhattan Project. In retrospect, it is a telling irony that the project that had spawned the danger of a final holocaust could, at the same time, stand as a model of the potential of science. Bateson wrote in the journal *Science*:

> We have not enough basic knowledge of the mechanics of individual aspiration and large-scale political interrelationships to plan the steps which must be taken to adjust human societies to the availability of atomic weapons. . . . It is possible that a small number of carefully selected men with experience in the modern handling of natural science problems might, after intensive training in psychological and anthropological methods, make outstanding contributions in the field of social science. (Bateson 1946)

Bateson, like Frank, was concerned with social lag and adjustment, but in the *Science* article these organicist themes had taken a technocratic turn. Selected scientists could be granted a special role: using modern scientific methods, they could provide the perspectives on society as a whole necessary for planning its development. Yet, if anything, these early systems theorists perceived their science to be anti-technocratic. The organicist view emphasized subservience of the individual to the larger group, or, at least, adjustment of the individual to a society composed of an integration of specialized groups. Feedback systems, in contrast, appeared to relieve the individual from total subservience to the function of the group—or to the decisions of the technocrat. Frank, speaking about the social sciences at the 1946 New York Academy meeting, described this possibility forcefully:

> Economists, political scientists, and sociologists . . . impute to the individual the motives and desires which they derive deductively from statistical studies of group activities . . . ignoring what the study of individual personalities has revealed about human behavior. . . . They will find it fruitful to recognize that the regularities of social life arise from the social-culture patterns and institutional practices into which the individual's activities are channeled. They will then realize that the dynamics of social life arise from individual actions, re-actions and interactions, not from mythical "forces" which they assume operate and control the social "system." (Frank 1948, 195)

(Frank's scare quotes around *system* here accentuate his rejection of social thought in which impersonal forces are viewed as responsible for the development of some social whole.)

Communication was a favored theme of the postwar Macy conferences on cybernetics. For Frank and other Macy participants, the individual in a feedback system appeared to gain in autonomy because systems theory addressed communication and information flow between individuals (Wiener 1948a, 1948b). Furthermore, the conferees could see a prospect of new social roles in scientifically managed systems. In contrast to domination of labor by paternalistic managers who served the owners of capital, social influence could be pluralized. The new managers, trained in cybernetic science, could institute more efficient systems of feedback—systems responsive to the needs of interacting individuals.

Nevertheless, appearances notwithstanding, the possibility of enhanced autonomy for the individual was problematic. Communication systems were also command-control systems; command-control engineers would be required to ensure that the systems operated according to new criteria; for example, minimizing information loss or preserving circuit stability. The technocratic implications of systems thinking were scarcely discussed and were perhaps invisible at this time to the Macy scientists. The implicit common endorsement of the systems engineering role by these people, who differed considerably in their attitudes toward politics, the military, and technology, revealed a shared wartime experience of social facilitation as scientists. (Wiener, who refused to undertake research for the military after

the war ended, stood as a possible exception.) The prospect of scientifically managed systems had transformed the discourse about science and
society; a technocracy, even though the term was not used, seemed both
realizable and acceptable. The ambiguous implications for individual freedom were obscured by a triumphant optimism: the social order that had tri-
umphed over fascism in the war was basically good. In fact, it was the best.
Technocratic optimism had supplanted organicism as the dominant mode
of social thought.

Systems as Allegory

In *Cybernetics,* Norbert Wiener cautioned against "excessive optimism"
about the "social efficacy" of cybernetics. From believing in the necessity of
control over the social environment, he said, "they come to believe it possible" (Wiener 1948a, 162). Making ecological and social engineering possible was precisely the task to which the young Odum applied himself.

The postwar alliance of government and science, and its attendant
optimism, provided fertile ground for the start of Odum's scientific career.
As I have described, he complemented the progressive organicism of his
father with the systems thinking of Hutchinson and the Macy conferees.
The ecology he built on those foundations was "macroscopic" (Odum
1971, 10). Using data derived from a system, the ecologist could stand
back and model the overall functioning of that system. Knowledge of the
system's structure was more important than its particular details for pre-
dicting the system's future behavior. The same approach could guide intervention in any system, whether its components were ecological, electrical,
or social.

However, as important as these conceptual developments were, they
constitute only one half of the story. Odum also derived support for his
early ecosystem research from the major institutions of postwar science—
the Office of Naval Research, the Atomic Energy Commission (AEC), the
Rockefeller Foundation, and the National Science Foundation (NSF). Such
funding for basic research in areas outside medicine and defense was rare
in the era before Sputnik. Science in the postwar decade was "big" only
where it intertwined with the military (Kargon 1983; Dickson 1984).
Moreover, Odum and his brother were among the few ecologists who
found themselves a place in this big science enterprise. E. P. Odum sur-
veyed the site of the Savannah River nuclear facility for the AEC, which led
in 1954 to a contract to study the coral reef ecosystem at Eniwetok Atoll, the
site of the Pacific H-bomb tests. H. T. Odum joined the Eniwetok project.

Other contracts followed, including NSF funds for studies of ecological microcosms constructed to investigate directly Odum's proposed techniques for ecological engineering (Odum 1962a).

Odum's project of designing "systems of man and nature" surfaced in the paper he wrote with his brother on the Eniwetok coral reef (Odum and Odum 1955). Its prominence in his ecological writings increased over time, culminating, in 1971, in his first book for a general audience, *Environment, Power, and Society*. This book reads in parts as an allegory for his style of doing science. Odum discussed a future in which power—that is, energy available per unit of time—might be expanding, but would more likely be constant or receding. He concluded that a

> bright possibility is ecological engineering. Adequate knowledge about the natural solar-energy based system [if power is limited] may allow a small concentrated loopback of energy to guide the systems of fields, forests and seas to stabilize and produce for man. . . . Loopback focus of relatively small energies may be able to hold world organization. . . . Greater work expenditures are required to create novel structure than to maintain it. The preparations need to be made now for this contingency of receding power. (H. T. Odum 1971, 309)

H. T. Odum, like Howard Scott before him, had a vision that reduced the complexity of social and ecological relations to a single energy dial for the social engineers to adjust. In his high-quality, low-energy circuits, Odum had found "in nature" a special role for systems engineers, working in the service of society.

Yet by this time, now twenty-five years after the end of World War II, an edge to Odum's optimism can be detected. He warned that, if power were receding, people needed to plan for the necessary transition. Odum had also turned his attention to the "disordering" effects of war, which drain energy away from "the maintenance of the structure of whole countries." After making a model of the effects of hurricanes on a mangrove swamp, he built an analogous model of the disordering—that is, war-losing—U.S. intervention in Vietnam (fig. 3.4; Odum et al. 1974). Although I have taken the two parts of figure 3.4 from different chapters of a later text (Odum 1982), the juxtaposition is appropriate. Odum and his co-workers prepared both diagrams following a visit to South Vietnam in 1970, as part of a National Academy of Sciences research project on the effects of defoliating herbicides. Moreover, the research team noted parallels between the two systems (Odum et al. 1974). A mangrove swamp, constantly reordering itself in response to decay and tidal action, could withstand the infrequent

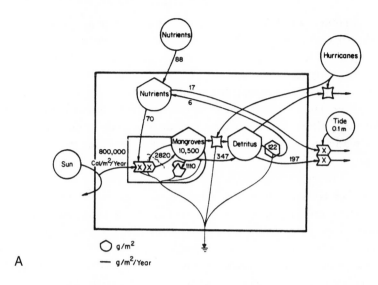

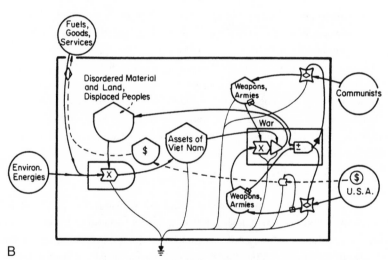

Figure 3.4 (A) Simulation model showing the effects of tides and hurricanes on a mangrove swamp. (B) Simulation model of the Vietnam war zone in 1970. Solid lines represent energy flows; dashed lines in B represent flows of money. The U.S. war effort was converted into energy at a rate of 14,000 calories per dollar. (From Odum 1982, figs. 21–37 and 25–27.)

input of disordering energies from hurricanes. The self-maintenance of the Vietnamese economy was, however, overwhelmed by the high level of the U.S. war effort. Consequently, "there was little doubt that the country would fall" (Brown 1977, 394).

War planners, and by extension, engineers in general could produce disastrous results. Odum hoped to make them recognize what they
overlooked. Society, like all systems, must ultimately submit to the maximum power principle—a natural law applicable to mangroves and
economies alike—a law people neglect at high cost. Odum, misunderstood
and unappreciated by the new systems ecologists, implied that there
existed a role above systems engineer in the enterprise of designing "systems of man and nature." At the close of his 1971 book he made this role
explicit:

> While there is energy, we need to simulate religious evolution. We may encour-
> age faster religious change even now by injecting large doses of systems science
> into the training of religious leaders. What a glorious flood of new revelation of
> truth God (the essence of network) has handed man in the twentieth century
> through sciences and other creative endeavors. How false are the prophets who
> refuse even to read about them and interpret the message to the flock. Why do
> some inhabitants of the church pulpits fight the new revelations simply because
> the temporary prophets are a million spiritually humble little people in laborato-
> ries and libraries, only vaguely aware of their role? Why not open the church
> doors to the new religion and use the preadapted cathedrals and best ethics of the
> old to include the new? Let us inject systems science in overdoses into the semi-
> naries and see what happens. Why should we fear that deviation from rigid sym-
> bols of the old religion is deviation from morality? A new and more powerful
> morality may emerge through the dedication of the millions of men who have
> faith in the new networks and endeavor zealously for them. Prophet where are
> thou? (H. T. Odum 1971, 310)

Clearly Odum saw himself as the prophet. Although his personal allegory is somewhat eccentric, it should be remembered that Odum was—and
continued to be—a successful scientist. The systematic social-personal-
scientific correlations I have discerned in his case illustrate how scientists
can build their experience of social relations and social action into rep-
utable science. In the research that pioneered the important features of systems ecology—in particular, its technocratic orientation—Odum was
making science of his place in society. His vision of designing "systems of
man and nature" gave him access to resources—institutional, technical, and
personal—for pursuing prodigious amounts of ecological research. Society
facilitated his work—in the broadest sense, and in ways not possible before
World War II.

B. Another Look: Diagrams and Physical Analogies

How can interpreters of science find significance in everyday practices
of scientists, such as drawing diagrams?

I have described strong social-personal-scientific correlations in the work
of H. T. Odum and explained them in terms of his life experiences as a scientist and the wider social concerns shaping his scientific theories. What
makes this causal linkage plausible are the facilitations Odum experienced
that enabled him to *act as if* ecology and society were composed of feedback systems—natural units that could be analyzed, theorized, and managed in terms of their energy flows. Social-personal-scientific correlations
that result in this way are not at odds with scientists' participation in dual
dialogues—with nature as well as with other scientists. To show this consistency, however, I could not focus on the correspondence between Odum's
theory and nature. I had to subsume the nature-theory dialogue within a
larger frame; namely, the interventions in nature and society, such as the
radioactive perturbation of the Puerto Rican rain forest, that Odum was
helping or hoping to construct.

I would not claim that Odum, with his visionary style, is typical of scientists, or typical even of scientists, such as Allee, whose social concerns are
evident in their science. In general, when social-personal-scientific correlations are evident, they probably need to be explained in terms of a more
diffuse mix of the different causal possibilities outlined at the start of this
chapter. Nevertheless, the image of scientist as agent-in-society that
emerged in my account of Odum should, in an important sense, be quite
familiar. He, like most people, sought cross-reinforcement among the overlapping realms he inhabited—the social, personal, and scientific—so that
efforts he made and directions he pursued in one realm did not undermine
those in the other realms. This image of Odum's agency—albeit a very general one in need of elaboration—renders it less remarkable that the life
experiences and social concerns of scientists can influence their answers to
scientific questions.

Odum is unexceptional in another important sense: he partook of many
of the mundane practices that constitute science everywhere. One of these
practices is making diagrams, which appear everywhere in Odum's publications. His extensive diagramming does not, however, on its own say anything interesting in relation to social-personal-scientific correlations. If
diagrams always duplicated propositions that were made in the text, they
would be merely illustrations. What would it take to show that diagrams

were more significant resources for Odum, that they contributed to the shape of his theories and his attempted reconstructions of society and nature?

In general, when scientists use diagrams to accompany texts and lectures, they make multiple references: to the phenomena overtly represented, to analogous phenomena or devices, and to previous pictures and their conventions of pictorial representation. When formulating diagrams, scientists have to delimit a phenomenon, choose and organize its components, and render on a flat surface three-dimensional entities and processes that change over time. They must do all this in a manner that facilitates the viewer's comprehension of the diagrams, illuminates their theory, and, more broadly, generates support for their scientific agenda.[13] In this second look at Odum's work, I show that all these aspects of pictorial representation placed limits on the physical analogies that Odum and other ecologists explored. Perhaps *limits* conveys too strong a sense of a judgment made in hindsight, for the analogies used actually facilitated Odum's work. His scientific activity—of which pictorial representation constitutes one strand—was richly metaphorical (notes 11 and 12). He was not only able to *think* that ecosystems are *like* feedback systems and energy circuits, but to *act as if* they were systems and circuits.

The Search for Physical Analogies

The Macy conferences popularized the view that complex situations of diverse kinds could be treated as self-regulating feedback or cybernetic systems (see section A of this chapter). The conferences drew on an optimism born of the experience of American wartime science, in which research teams, well funded and organized by the government and the military, had produced enormous scientific and engineering advances. Conference participants believed that the successes of the physical sciences could be extended to the biological and social sciences. Long-time runs of data would enable systems scientists to detect feedback and expose the dynamics of systems. Cybernetic social science promised increased scope for scientific management of society, although no one was explicit about who the engineers of social systems would be. Finally, a general theory of cybernetics offered common terms that would apply to all feedback systems, whatever their mix of living and nonliving components. Toward this unification, concepts were actively borrowed by one science from another (Heims 1991). I shall show all these themes in the ecosystem diagrams that follow.

Ecology was among the sciences seeking concepts, methods, and theoretical formulations in the physical sciences. When Hutchinson described his "biogeochemical" approach for analyzing "circular causal systems in ecology" (Hutchinson 1948), he illustrated it with a diagram depicting the budget of carbon (C) in the biosphere (see fig. 3.1). The diagram combined standard notation for chemical reactions with arrows corresponding to very general pathways of the flow of carbon, particularly in energetic transformations such as "animal respiration." The structure of this diagram, while novel for ecology at that time, was common in biochemistry, in which metabolic pathways involved in, for example, fat metabolism were summarized using arrows linked into cycles. Paired arrows indicated reversible reactions; dashed arrows highlighted less well-established pathways (see, e.g., Baldwin 1947, 413). The borrowing of this graphic vocabulary was not incidental; it facilitated biologists' reading of the diagram at the same time that it conveyed Hutchinson's idea that ecological processes were a subset of interlocking chemical processes.

Odum's dissertation under Hutchinson employed two specific analogies to extend the idea of a steady state cybernetic mechanism. One followed Alfred Lotka's exposition of "physical biology" as analogous to physical chemistry (Lotka 1925); the other followed the ecologist George Clarke's 1946 depiction of a marine food web as a set of cogwheels (fig. 3.5). Odum had clearly not settled on a precise physical analogy appropriate for theorizing ecological complexity. In fact, it was neither Lotka's nor Clarke's analogy that he developed in his next research project, but rather Lindeman's program of measuring energy stocks and flows in the trophic structure of an ecosystem. Nevertheless, a common message emerged from the analogizing: ecology could aspire to the theoretical status of physical sciences.

Before I move to Odum's early ecosystem diagrams, let me examine Clarke's cogwheel diagram in more detail, as it reveals issues glossed over in Odum's reference to "cogwheel[s] . . . geared to other cogwheels" (Odum 1950, 308). In employing diagrams, Clarke had to solve expository problems, to figure out a way to express his ideas. Clarke's article concerned two sets of distinctions within the "production cycle" in a marine environment: (1) the subdivision of gross production at each trophic level into parts respired, consumed directly by other organisms, added as net growth, and left to decompose; and (2) the standing stock versus the production rate at the different trophic levels. With the cogwheel diagram, he left aside his first distinction. The appeal of cogwheels to Clarke was that the body of the cogwheel could represent the standing stock, and rates of net production could be indicated by "the rates at which the cogwheels are

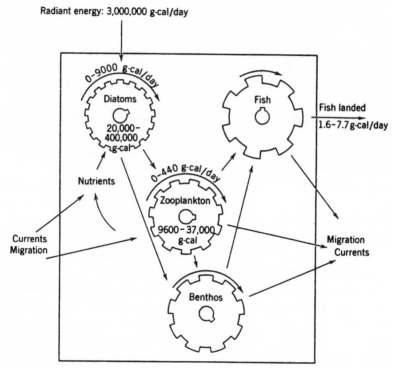

Radiant energy: 3,000,000 g-cal/day

Figure 3.5 Components of the Georges Bank ecosystem. The body of each cogwheel represents the standing stock; quantities are the observed maximum and minimum values per m² of sea surface. The longer the arrow curving around a cogwheel, the greater the production rate. (Reproduced from Clarke 1954, 499, but originally printed in Clarke 1946, 333.)

turning" (Clarke 1946, 333). The longer arrow curving around the diatom cogwheel, for example, represented diatoms' greater production rate. Such a comparison necessitated a common unit of production; therefore, everything was converted to its energetic equivalent (e.g., 1 gram of fish = 740 g.calories, i.e., 3100 joules).

Like all analogies, physical analogies made in diagrams have their limits. Clarke's diagrams dealt poorly with the scale implied by his subject. The standing stock for each trophic level was given on the face of the cogwheel. Although diatoms also have a much larger standing stock than the other trophic levels, Clarke decided not to scale the cogwheels accordingly. The only indications of scale were the number and size of the teeth. The cogwheel for diatoms—microscopic organisms, numerous and producing at a fast rate—had many small teeth; for fish—larger and slowly producing—fewer, larger teeth were depicted. But note that the gearing was then incorrect. (Although the teeth were not interlocking, the implication of gearing was clear.) A wheel with more teeth turns more slowly than a wheel with fewer teeth geared to it. If Clarke was aware of this problem, he gave no sign of it, and chose to reproduce the diagram in his 1954 textbook and its 1965 reprint.

Similar expository problems may yield quite different graphic solutions. Several years later, when Odum reported on the measurements from his first study of energy transformations up different trophic levels in the Silver Springs of central Florida (Odum 1956b), he also addressed the problem of depicting standing stock and production rate. His diagram conveyed the relative scale of both the biomass stocks and the production rates (again, all converted to energy equivalents) and also addressed Clarke's other distinction, the subdivision of gross production into respiration, consumption, and so forth (see fig. 3.2). While the form of this diagram was new in ecology, its flow of energy played off a familiar "hydraulic" theme used, for example, in popular literature to depict the functions of the body, such as the digestive system.

Although Odum has described the step to energy flow diagrams as obvious for someone with his biogeochemical training (Odum 1986), his diagrams had a proportionality and clear orientation that Hutchinson's biogeochemical diagrams lacked. The flow downward, from the upper left to the heat sink at the bottom of figure 3.2, adds the force of gravity to Odum's dominant theme, that ecological and biogeochemical cycles are "driven by radiant energy" (Odum 1951, 407).

While serving economically to summarize measurements, an energy flow diagram suggests little about how the ecological processes measured might be theorized. Hutchinson and Lindeman had hinted at the possibility of a mathematical analysis of ecological dynamics when they used λ to denote the relative production rate at successive trophic levels (Lindeman 1942; Hutchinson 1942). In a similar diagram (Odum 1956a, figure 4), Odum employed and added to this mathematical notation, but he was far more theoretically explicit. In his major paper on the Silver Springs, he produced eight hypotheses for investigation concerning biological communities and their partitioning of energy (Odum 1955).

Odum's theoretical productivity accelerated in the late 1950s when he made the step of converting his energy flow diagrams to actual electrical circuits (Odum 1960). Figure 3.3, which depicts such an electrical circuit, corresponds very closely to figure 3.2. The boxes corresponding to the trophic compartments are similarly placed and scaled. Inside the boxes in figure 3.3, variable resistors are represented by the zigzag lines with arrows. In the electrical circuits that Odum built and manipulated, the resistors could be adjusted until the current flow along the connecting wires, measured on the many ammeters, became proportional to the measurements of energy flowing between the compartments in the actual ecosystem. (Since electrons cycle in figure 3.3, but energy is eventually dissipated in figure 3.2, figure 3.3 is strictly analogous to the flow of carbon that accompanies the energy transformations in figure 3.2.)

The two diagrams depict in a similar way the relative sizes of trophic compartments. In the circuit, however, there is no material analog to the biomass, and so the boxes in the diagram are superfluous in a functioning circuit. Yet the boxes add weight to the diagram as a bridge between the ecosystem and the electrical circuit—the circuit was a physical device that Odum could use with facility to deliver on the theory only promised by earlier diagrams. While other pioneering systems ecologists were beginning to use analog and then digital computers to simulate equations governing energy and material flows (Bocking 1997), Odum insisted on working with material systems. This direct analogizing led Odum to introduce storage condensers into his electrical trophic compartments, making the charge stored in each condenser equivalent to the biomass. Reciprocally, he sought an ecological equivalent for the voltage drop between compartments and suggested the term "ecoforce," measured by the voltmeter in figure 3.3. This first exploration of ecosystem circuits yielded fifteen more hypotheses for investigation.

In a 1961 conference on mathematical biology, Odum made his method of analogizing explicit: "One asks two kinds of questions: (1) What is the electrical significance of a function observed in nature; or (2) given an electrical unit in a circuit, what is it in the ecological system" (Odum 1962b, 295). Relying on the "exact energetic equivalence" of the ecosystem and the electrical circuit, ecological theory could advance rapidly by borrowing from the better-developed electrical theory. He believed that eventually ecology might "feedback into the physical sciences, because ecological systems have the power to self-regulate and self-maintain, something not yet achieved in physical science" (Odum 1962b, 291–92).

Graphic Conventions

This interpretation of ecosystem diagrams has thus far centered on the primary ideas that three ecologists summarized in them. The diagrams, however, represent more than their obvious referents, as will become evident when I examine some of the graphic conventions employed.

Framing: Diagrams for scientific publication must be inserted on a flat and rectangular printed page in some relation to the text. In earlier days, when it was common to print illustrations on pages or plates separate from the text and when several elements were arranged on a single page, they were often enclosed within a frame (Blum 1993). Clarke's cogwheels and Odum's electrical circuit preserve the frame, but in both cases extend the convention to demarcate an inside from an outside. The internal components are separated from external factors. Materials and energy flow across the frame or boundary, but external factors impinge in a simple way on the coherent internal relations. For example, the interconnected biotic cogwheels dominate Clarke's marine food web diagram, and the light energy enters through a simple arrow.

Such partitioning into inside and outside gives weight to the idea that nature can be divided into systems. The frame around Odum's circuit diagram is particularly strong. At the material level, it depicts the well-defined edge of his circuit board. But, at the same time, the framing convention conveys the system-ness of ecological relations and suggests that an analyst can observe the system as a whole from the outside. In an actual ecological situation, it would be difficult for anyone to find a position in nature providing such an overview. Electrical circuits, on the other hand, can readily be analyzed and controlled to achieve some overall function or output. The cybernetics ideal of managing complex systems—Odum's "systems of man and nature"—has been central since the very early days of his ecological research (section A).

The nature of the printed page ensures that partitioning into inside and outside will be an issue even if no explicit frame is depicted. Although diagrams can be drawn as if they continued off the page, only the advent of some computerized scan and zoom display would minimize the need for the diagrammer to make decisions about what to include and exclude. Without such a display, the very act of making ecosystem diagrams on printed pages favors a system view of nature. Moreover, when diagrams depict ecological situations of widely varying scale and extent, an even stronger theoretical claim is implied; namely, that ecological complexity can be partitioned into a hierarchy of systems within systems. Just as one usually overlooks the graphic limitations of the printed page, many ecologists take the partitionability of nature as an unquestioned precondition for ecological experiment and theory (Taylor 1992a).

Space and time: Pictorial conventions are also essential in reducing objects and processes to the two dimensions of the printed page. Shading and contours create the illusion of three dimensions. Time may be conveyed by proxy through a spatial depiction of a narrative, as in Minard's well-known map of Napoleon's advance to Russia and subsequent retreat in 1812 (see Tufte 1983). In Odum's ecosystem diagrams, however, space and time remain virtually unrepresented. The separation of trophic compartments is not intended to connote the spatial location of the plants, herbivores, and so on. The flows occur in time, but are constant. Like all equilibrial formulations, they show no history or prospect for change. The only spatial reference, besides the sunlight entering at an angle from above, is to inside and outside. The whole system is inside the frame, and each trophic compartment is contained within a box.

If attention were paid to the spatial or temporal arrangements within the ecosystem, the living activities concealed behind the measured energy flows would be brought back into the picture. Their absence in the diagrams favors particular kinds of ecological analysis and theory. Odum's diagrams imply that the activities of different organisms can be measured in terms of energy captured and expended. According to both his claims and his diagrams, organisms can be unambiguously separated into trophic levels and aggregated for the purpose of understanding ecosystem function. Furthermore, the structure and function of the ecosystem are depicted as stable. Finally, as discussed above, an ecosystem having an inside and an outside suggests a location or role for the external analyst, who maintains an appropriate scientific distance or managerial overview. Notice that, in contrast to an ecosystem, an electrical circuit is made up of components that are physically localized and can be arranged as in a flat diagram. The separateness of components, well represented in a circuit diagram, is characteristic of engineered systems, but not of natural ones.

The preceding interpretations were not, of course, formulated only by examining the diagrams. My view has been informed by the accompanying texts and by the larger context of Odum's work (section A). Nevertheless, the diagrams are by no means redundant for the interpretative process, just as they have not been incidental in Odum's work. The same information contained in an energy flow diagram could have been represented differently; for example, as in figure 3.6 (based on Clarke's marine food web). Like Odum's energy flow or circuit diagrams, this diagram highlights the centrality of measurement to systems ecology. Yet it provides no sign or promise of a physical theory to explain the energy data. Further, the viewer or systems analyst is left without a privileged vantage point on the ecological situation.

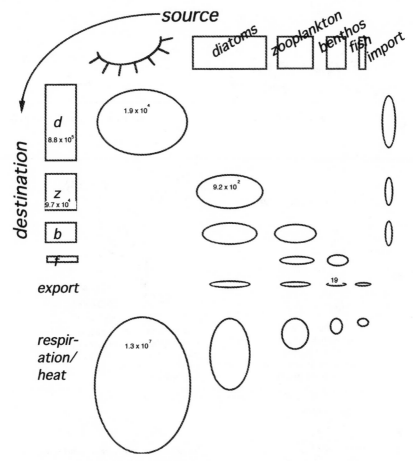

Figure 3.6 A matrix representation of energy flow, using the same data as in figure 3.5. The sizes of the ellipses are proportional to the energy flow from each source to each destination. (From Taylor and Blum 1991a.)

The Search for Universality

While analogies to physical systems promised or provided theory for a science of ecological systems, they also represented Odum's aspirations for universality of that theory. When Odum introduced his maximum power principle and applied the same general formulation to many situations, a diagram of similar form accompanied each case (Odum and Pinkerton 1955). Four of these diagrams are juxtaposed in figure 3.7. In the years that followed, Odum explored flow diagrams for systems of widely varying

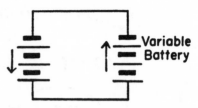

FIG. 4. Water wheel driving a grind-stone.

FIG. 5. One battery charging another battery.

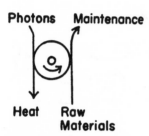

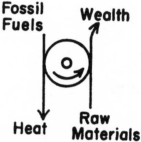

FIG. 11. Climax community. The absorption of photons drives the processes of growth necessary to maintain and repair the community.

FIG. 12. Growth and maintenance of a civilization. The use of fossil fuels in power plants, etc., drives the production of wealth both for maintenance and growth.

Figure 3.7 "Schematic diagrams" of a selection of physical, biological, and social systems. (From Odum and Pinkerton 1955 [captions are theirs].)

scale and kind, defining and arranging the compartments so that when the diagrams were presented together, structural similarities would be evident. His comparative use of energy flow diagrams showed clearly a commitment to the unification of science. It was the ecological circuit diagrams, however, that unleashed Odum's universalist energies. If electrical system theory was to be borrowed by ecosystem ecology, and vice versa, a common language could facilitate this communication.

During the 1960s, while Odum was leading the Puerto Rico rain forest project, he developed and refined a set of symbols for systems based on his energy circuit diagrams. In the early stages of promoting his universal system language, the new energy circuit symbols were introduced together with standard and more familiar electrical symbols (fig. 3.8). Conceptual shifts, some evident in figure 3.8, further facilitated the universalist project. Notice that in the energy circuit diagram (fig. 3.8A), the detail of circuitry for the biomass compartments was simplified into the hexagonal and capped semicircle symbols. This paring away of detail left the structure of the compartments and their interconnections in bolder relief. Notice also that, although the compartments were still loosely scaled to the biomass, the flows of energy were no longer conveyed with any sense of proportionality or measure. In fact, although some data were superimposed on figure 3.8, Odum drew thousands of diagrams of nonecological systems for which he had no data to present. All that remained in these diagrams was the structure. If diagrams without measurement were to support his universalist aspirations, then system structure had to be the key to understanding systems. Reciprocally, if system structure were the key to theorizing about systems, then diagramming should be a key ecological method. And, indeed,

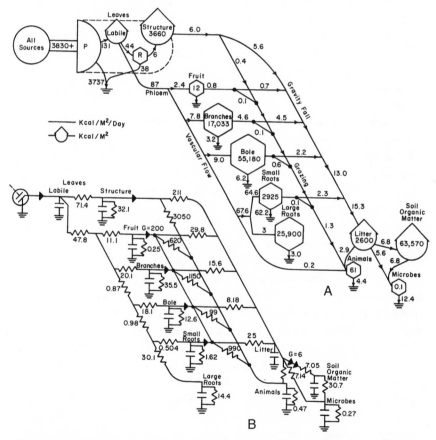

Figure 3.8 "A network energy diagram of the main power flows [in a tropical rain forest in El Verde, Puerto Rico]. No feedback from storage to power intake is shown within or between compartments. (A) Energy-circuit language. (B) Equivalent-circuit language." (From Odum 1970, p. 1193.)

when Odum subsequently compared systems of widely varying kinds (see, e.g., fig. 3.4), energy circuit diagrams became the hallmark of his publications. When his synthetic text, *Systems Ecology* (Odum 1982), appeared, its 580 pages contained about 1,400 energy circuit diagrams.

Although energy circuit diagrams and a theoretical emphasis on system structure strongly reinforced one another, there was another dimension to Odum's universalist project. Other cyberneticists and ecosystem theorists explored different measures of system structure, usually based on information theory (see Ulanowicz 1986), but Odum insisted on energy as his currency. Information and monetary flows were simply equivalent to high-quality energy flows (Odum 1982, 12, 18–19). Energy provided a link to real, nonabstract physical systems and also, via the maximum power principle, to the power and status of thermodynamic principles. With these underpinnings, energy circuit analysis became a natural means of reducing complexity. A study of the self-maintaining operation of a mangrove swamp, for example, could inform interventions in a war economy (see fig. 3.4).

Unlike most of Odum's energy circuit diagrams, the two in figure 3.4 preserve the frame noted earlier: hurricanes, communists, and the U.S.A. are external influences on the mangrove or war zone systems. The frame suggests the possibility of system engineering; indeed, Odum and his collaborators concluded that the war planners could have better managed their inputs into South Vietnam (Brown 1977). Moreover, the diagramming obscures all other potential agents. Individual people, like other organisms, are aggregated into compartments; the detail of their activities is lost through energy accounting.

In the "macroscopic view" of *Environment, Power, and Society* (Odum 1971), the elimination of detail is a positive step. "Men already having a clear view of the parts in their fantastically complex detail, must somehow get away, rise above, step back, group parts, simplify concepts, interpose frosted glass, and thus somehow see the big patterns" (Odum 1971, 10). The language here is universalist and the intended audience broad, yet in Odum's proposed social organization, the proponents would hold special places—just as they do in all technocratic visions. It was systems engineers for whom ecological complexity was grouped, flattened, and framed into circuit diagrams. These diagrams were, in turn, assembled by a prophet of ecological systems engineering. Odum combined the diagrams with innumerable unanswered but leading questions (see also Odum 1982, 582), which seemed meant to inspire the inquiring and concerned to see nature and society as the prophet depicted them.

Coda

In the 1970s and 1980s, Odum attracted an international following of environmentalists, energy policy analysts, and economists (Hall 1995). These researchers and educators shared Odum's vision of "systems of man and nature," designed and managed according to a natural reduction of complexity. However, the main trajectories of ecological theory led in other directions. This shift may, in part, have reflected the problems ecologists have reading dynamics from patterns in data (see note 10) or finding systems that are in equilibrium (see chapter 1, section A, and note 8). Some ecologists made a move that took them away not only from system analyses, but also from early community ecology. Following the lead of evolutionary biologists, they shifted their focus to the idea that individuals pursue their own self-interest, turning their attention away from discussion of how groups or systems might evolve toward improved feedback and coordination. Ecological organization—to the extent that these ecologists acknowledged its existence—was theorized as an indirect consequence of the more fundamental processes through which individuals colonize space, capture energy, grow, survive, disperse, and so on (Dirzo and Sarukhán 1984; Givnish 1986; Taylor 1992a). (An equivalent shift occurred in explaining cooperation and social organization; for historical interpretations, see Haraway 1981–1982; Keller 1988.)

When compared with the immediate postwar context, the social and practical conditions also became less conducive to acting as if entire ecological systems could be analyzed and managed (Bocking 1997, 179–205). At a very broad social level, the shifts in predominant patterns of thought spoke both to the postwar fulfillment of technocratic optimism and to its subsequent eclipse. A diversity of roles for scientists in social management had indeed opened up since World War II. At the same time, the influence of any single scientist became diluted; fewer scientists could identify as closely with social leadership as could H. T. Odum or H. W. Odum before him. In society more generally, the individual's ties of personal dependence within the family and community were loosened, and the individual became freer to consume and seek personal fulfillment. At the same time, the individual became more vulnerable to decisions about social organization that were increasingly made on a scale that eliminates the details of local interests and at a distance that removes the decisions from the individual's influence. In fact, in feedback systems made of energy or information, the individual is potentially obsolete; all processes—physical or biological—are substitutable. The exercise of a technocratic rationality that Odum helped naturalize diminished the grounds for feeling secure or optimistic about one's place in the changing world.

At a more specific level, the prospect of whole-system analysis and management diminished, paradoxically, as environmental planning improved in the 1970s. The institutionalization of environmental impact assessment led to increased scrutiny of the conclusions of environmental scientists, especially during adversarial judicial proceedings. These assessments, moreover, had to be prepared in less time and with fewer resources than systems analyses usually required. For example, during the 1970s, studies of the impact of Hudson River power plants focused on one part of the ecosystem—the populations of striped bass—because this part was of most concern to the competing parties. For this task, the "simulation modeling of the river environment was less suitable than other methods relying on local empirical evidence" (Bocking 1997, 109).

In general, the practices of environmental assessment have been governed more by complex and contested pragmatics than by all-encompassing theoretical programs of ecologists such as Odum, his brother, and the leaders of the International Biological Program (Bocking 1997; see also Edwards 2005; Jacobson et al. 2000; and Taylor 1997c on the rise of earth system science in the context of anthropogenic climate change). Accordingly, one might expect social-personal-scientific correlations to show less consistency over time and to be less obvious than in Odum's case. Exposing and interpreting less direct relationships between scientists' representations and actions is a challenge I take up in the next chapter.

4

Reconstructing Heterogeneous Webs
in Socio-Environmental Research

After considering H. T. Odum's practice—his methods and organization of research as well as his concepts and production of theory—I saw him as a person or agent working to make the overlapping realms he inhabited—the social, personal, and scientific—reinforce one another, so that efforts made and directions pursued in one realm did not undermine those in the others. In my interpretation, many aspects of the postwar setting for Odum's early research enabled him not only to think that ecosystems were like well-designed feedback systems (or circuits), but also to act as if they were—he was able to find in nature a special role for systems engineers, such as himself, working in the service of society.

Yet I wondered about the generality of this model of scientific and social agency. Could I show reinforcement across realms in cases in which the social-personal-scientific correlations were less obvious or less consistent over time than in the case of Odum's scientific work? In this line of inquiry about scientists as social agents, I followed the lead of sociologists of science, especially sociologists of scientific knowledge, who had been formulating vocabulary and propositions about how scientists in practice establish knowledge (Collins 1981a).

Another challenge remained after interpreting Odum's work. I had shown that the sociality of science could affect the content of scientific knowledge, but my original motivation was to bring such interpretations to bear productively on subsequent research. In this regard, the case of Odum provided limited guidance. Personal, scientific, and social considerations reinforced one another so consistently in Odum's life and work that it was difficult to see how he could have done anything differently. At best, I could have used my interpretation of Odum to suggest a very broad lesson: Scientists opposed to technocratic rationality should not treat ecological complexity as if it were made up of well-bounded systems that could be analyzed in terms of a single currency. Yet, any scientists who wanted to heed such a lesson would still need specific ways to arrange or alter the personal, scientific, and social facilitations of their work. To provide insights about how that might be achieved, a finer-grained analysis than the broad historical interpretation of Odum seemed to be called for.

With these two challenges in mind, I chose to consider two projects of socio-environmental assessment likely to be governed by more complex and contested pragmatics. The first case was the modeling work I had undertaken in Australia, as part of a project analyzing the future of a salt-affected agricultural region (section A). The second involved U.S. researchers in the mid-1970s building computer models of nomadic pastoralists (livestock herders) in drought-stricken sub-Saharan Africa (section B) …

◆ ◆ ◆

A. The Simulated Future of a Salt-Affected Agricultural Region

How do research sponsors manage to get the results they pay for—could the results have turned out any other way?

In the Kerang region of the southern Australian state of Victoria, farmers irrigate some crops as well as pasture, which is grazed by beef or dairy cattle and sheep. Soil salinization has been a chronic problem; during the mid-1970s, after some very wet years, the problem was acute. The rise in salinity, following a decline in beef prices, threatened the economic viability of the region. In late 1977, the ministry of the state government overseeing water resource issues commissioned the "Institute," an economic and social research organization in Melbourne, the capital city 240 kilometers south of the Kerang region, to study the region's economic future. An agricultural economist from the Ministry and the principal investigator from the Institute formulated a project to evaluate different government policies, such as funding regional drainage systems, reallocating water rights, and raising water charges. This evaluation would take into account possible changes in the mix of farm enterprises and in farming practices, such as improvements in irrigation layout, drainage, and water management. The analysis was to be repeated for different macroeconomic scenarios as projected by the Institute's national forecasting models.

The central part of the project was the construction of what came to be known as the Kerang Farm Model (KFM) (fig. 4.1). Using a technique called linear programming for each of four composite representative farms, the KFM would determine the mix of farming activities that produced the most income. Different factors, such as water allocation, could be changed and the effect on the optimal income and mix of activities ascertained. The division of labor in the project was as follows: The principal investigator, an

Production

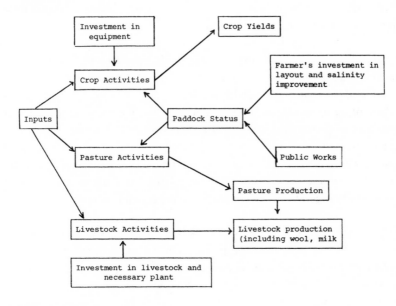

Economics

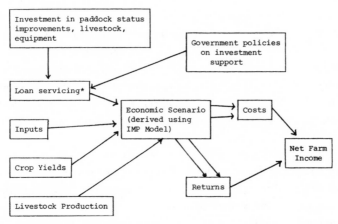

Figure 4.1 A schematic of the Kerang Farm Model. The arrows correspond to equations that capture the effects of parameters and variables (represented by the boxes) on other variables. (From Taylor 1979.)

econometrician, continued his work on the agricultural component of the Institute's forecasting model. An agricultural economist conducted extensive surveys of farm operations on forty farms and acted as liaison with two senior agricultural extension officers in the region who helped to refine the production relationships and parameters used in the KFM. I was hired for fifteen months as a statistician and modeler to analyze the farm surveys and to construct and operate the KFM. The Ministry maintained oversight of the project through its agricultural economist and through regular meetings with the project team and an advisory committee.

The study resulted in a number of tangible products: the survey and data analysis incorporated into one report to the Ministry; the KFM and economic analysis that made up the second report; a technical monograph documenting the KFM; papers presented at two national conferences of agricultural economists; and a public meeting in the Kerang region to explain the results of the study (Ferguson, Smith, and Taylor 1978, 1979; Taylor 1979). Even without refinements that were omitted to meet the Ministry's deadline, detail in the KFM was sufficient to allow evaluation of the required range of factors—the model was complex, yet still manageable to use.

At the public meeting to present the study's findings, some local agricultural extension officers objected to a conclusion that irrigation of pasture gave a better return than irrigation of crops. This conclusion ran contrary to the advice they had been giving to farmers ever since the decline in beef prices. Reanalysis, incorporating generous increases in crop yields into the KFM's parameters, was completed rapidly. It showed that the benefit of pasture irrigation was a robust result, which could be attributed to the recovery of beef prices by the late 1970s. The Ministry, meanwhile, focused its attention on results that indicated that water charges were not a primary limiting factor on farm enterprises or viability. Although the Institute had been commissioned to analyze a larger range of options, the other analyses were eclipsed by the conclusion about water charges, which suggests that justifying an increase in water charges had been the Ministry's primary concern all along.

Readers who have performed contract research on government policy may be all too familiar with the experience of results being used for purposes more limited than, or in a different spirit than, they had hoped. Cynics would assert that distortion is what one should expect when governments commission research—why should it turn out any other way? A less cynical, but still fatalistic, sentiment often expressed by researchers is that the best they can do is to produce results that are as faithful as possible to reality and hope that, eventually, the truth in these analyses will filter through the political process and act as a check on unjustified policy.

An alternative to cynicism and fatalism rests on teasing open the relationship between the scientific research and the circumstances in which it is conducted. If particular aspects of the relationship can be shown to influence the results in their own way, then one has also shown specific ways in which the research could have been carried out differently. Researchers need not wait, then, for the inevitable distortion of results or for the eventual acceptance of the truth; they can instead attempt to change the particular aspects over which they have the most influence. The possibility of identifying sites where research might be modified motivates the analysis to follow.

My analysis of the Kerang project begins with the modeling because that was the part that I, as a participant, observed most closely. I refer to myself in the third person as "the modeler" to express some distance between my position and actions in 1978–1979 and my subsequent interpretive role. Although I do not discount my observations and understandings as a participant (see Collins 1984 for a discussion of "participant comprehension"), it would be misleading to imply that during the Kerang study I had in mind the sociological analysis or themes about knowledge-making that follow.

Building and Probing the Kerang Farm Model

My analysis begins with the observation that the development of the KFM drew on many diverse "components" (fig. 4.2). (The reason for choosing this neutral term will emerge in due course.) These components included data on soil quality; expected crop yields; the range of farm sizes; the technical assumptions used in the linear program; the status of the different agents in the project; the geographic distance between the Institute and the Kerang region; the computer packages available; the terms of reference set by the Ministry; and so on. Moreover, many of these components link the different domains of social action, or "social worlds" (Clarke 1990, 1991; Fujimura 1988), of the various agents—from the modeler to the farmers. I need to put some order into this heterogeneity of components and assess their relative importance to the knowledge-making. My experience as the modeler allows me to unpack parts of the processes of model building here.

Consider the central technical feature of the KFM: the use of a linear program for economic analysis. This required the assumption that farmers operate to maximize one objective; in the KFM, this objective was income. Furthermore, the use of a linear program for policy formation assumed that if the optimal mix of farming activities according to the KFM were different from a farmer's existing mix, the farmer would change accordingly and

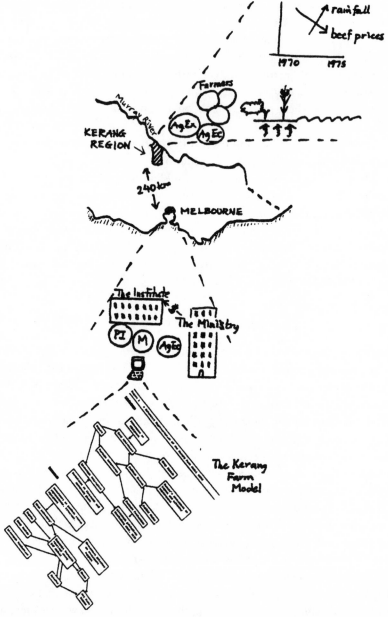

Figure 4.2 A schema of the diverse components involved in the production of the Kerang Farm Model. Symbols: PI, principal investigator; M, modeler; AgEc, agricultural economist; AgEx, agricultural extension officers.

immediately. Even though the economic future of the region obviously entailed the farmers' participation, the study was not designed to investigate why and how farmers change, how directly and readily they respond to economic signs, or the extent to which any overriding economic rationality governed their actions. Finally, the use of the linear program assumed that all the relevant activities could be well specified. However, the potential economic and ecological benefits of novel long-term options, such as selective reforestation and organic soil restoration, were difficult, if not impossible, to estimate without knowing the outcome of measures that had not yet been implemented, such as experimental plots, publicity, education, advocacy, and subsided loans for tree planting.

The modeler questioned these technical assumptions. He expressed interest in techniques that incorporated more than one objective, but the principal investigator could not envisage modeling an alternative objective to income. In any case, software for multi-objective analysis was not available at the computer center used by the Institute. The modeler designed the KFM to allow the effect over time of new investments to be examined, but when the project approached its deadline, this part of the model development was halted. The modeler learned of a sociological study on the factors influencing Kerang farmers to change their practices. This study had not, however, been released at that time, and the principal investigator lent no institutional support to obtain advance access to it. Finally, the modeler located informants with experience in reforestation and organic soil restoration, but was told that these and other issues were outside the economic specialization of the Institute and best left for others to deal with.

In affirming the technical assumptions in the KFM, the principal investigator invoked his senior and permanent position at the Institute, the terms of reference and deadlines that the Ministry had set, and the Institute's specialization in quantitative economic research. These assumptions, in turn, had several consequences for the research. They eliminated certain issues from investigation (e.g., farmer's views of their objectives). They shaped the data that needed to be collected (e.g., there was no need to investigate how farmers change or build scenarios for the novel long-term options). And they colored the relationships put into the model (e.g., the time course of investment became a secondary issue to the farmers' income-optimizing activities). The authority over the young modeler exercised here by the experienced principal investigator was not extraordinary. Nevertheless, through such exchanges, the principal investigator and the modeler were negotiating the different components of what would count as a representation of reality and a guide to policy formation.

Of course, there were parties other than the principal investigator and the modeler who might have been involved in accepting or disputing the KFM. The farmers might have objected to the way their behavior was modeled. The KFM could also have been disputed by economists interested in multi-objective techniques; by sociologists interested in how people act, interact, and change; or by agricultural policymakers interested in using the study to help change the state of farming in the region. None of these potential disputes proved significant at the time. The farmers were separated from the formulation and operation of the KFM. Conversely, the KFM was insulated from the farmers by several considerations: location (the modeling was performed in the city), the chain of personnel involved (modeler–agricultural economist–senior agricultural extension officers–local agricultural extension officers–farmers), and levels of abstraction and generalization. No one in the Institute, the principal investigator in particular, had training in multi-objective economic analysis or ready access to suitable computer software. There were no sociologists included in the project team or advisory committee. The Ministry, through the range of options established in the terms of reference for the study, indicated that change would be initiated by government policy based on economic and engineering criteria. The farmers were, in effect, to be instruments, more than co-participants, in determining the future of the region. In short, the Ministry did not dispute the KFM as a representation of reality; neither were any farmers, economists, or sociologists in a position to do so.

Six Themes Drawn from the Reconstruction of the Kerang Study

The description of the building of the KFM, although brief and partial, is sufficient to introduce six themes or propositions about the processes of making science and the interpretation of those processes. These themes are put forward in the spirit of theoretical exploration; that is, to highlight important issues and orient our thinking about them. I begin with the observation that the different agents involved in the KFM drew on diverse components from a range of domains of social action (theme 1). Themes about scientists as imaginative agents who represent-engage (themes 2 and 3) extend the idea of social-personal-scientific correlations used in chapter 3's analysis of Odum's work. The other themes take further steps toward a more general framework for analyzing the mutual relationships between scientists' representations and actions. Section B of this chapter makes these themes more concrete by applying them to another case study.

Theme 1. Science-in-the-making depends on heterogeneous webs, not unitary correspondences. From the description above, it is clear that diverse components were involved in the building of the KFM—from soil quality data to the terms of reference set by the Ministry. Moreover, they were interconnected in practice, forming what I will call *heterogeneous webs* (fig. 4.3). The KFM's assumption that farmers were subordinate to economic rationality made it easier to concentrate only on options that took the form of government policy. The power of the government to enact its decisions rendered investigation of how farmers change less relevant, which shaped the data that needed to be collected—generalized agronomic data, rather than sociological interviews, would suffice. This choice of data, in turn, conditioned the relationships that could appear in the model. Similarly, the modeler's mediated relationship with the modeled situation and his geographic separation from the region rendered it less relevant to model the novel long-term options. Their omission from the modeling, in turn, helped to ensure that such aspects of the future reality would be less realizable and the model's account more real.

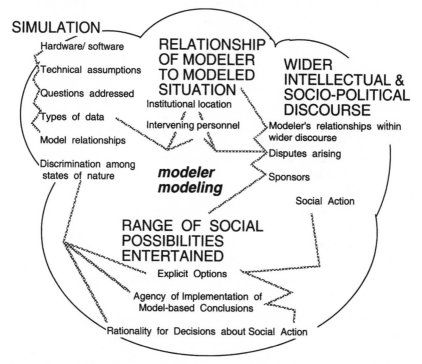

Figure 4.3 A schematic picture of the web of diverse, interconnected components involved in the building of the Kerang Farm Model. (From Taylor 1995a.)

"Technical" considerations, such as the assumption of income optimization, and "social" considerations, such as the separation of the modeler from the farmers, had implications for each other in practice. "Local" interactions were connected with activities at a distance. For example, the modeler and the principal investigator decided not to pursue sociological inquiry into how farmers change, which meant that the content of and conduct of the survey of farms and farmers could remain unchanged. No one component in the web stood alone in supporting the KFM as a representation of reality; in the actual process of building the model, technical components could not be detached from social ones, nor local ones from those that spanned levels.

In this sense, I would say that science is *constructed*: science-in-the-making is an ongoing process of building from diverse components, just as a house is built over time using plans and measurements, laborers and contracts, concrete and concrete mixers, wood and saws. This is *social* construction, but not *merely* social construction—my interpretation is not that scientific knowledge is determined by or reflects the society in which it is made. Although it is possible to say that the model *reflected* all the different social components, that would be stretching the metaphor of reflection. Given the heterogeneity of components and their interlinkage in an ongoing process, it is difficult and uninformative to collapse science-in-the-making into a unitary idea that scientific knowledge reflects society. Likewise for the unitary idea that scientific knowledge reflects natural reality. Science, in practice, is *heterogeneously constructed*.[14,15]

Theme 2. **Scientists represent-engage.** In the process of building the model, the modeler, principal investigator, and other agents linked together technical and social components in order to make a model that worked for them. These scientific agents tended to make the different components reinforce, not undermine, one another, rendering both the model and the ongoing scientific activity more difficult for others to oppose or modify *in practice* (see theme 1). This insight goes beyond observing that representations of natural reality support engagements or interventions in different domains of society (Keller 1992, 74ff), or claiming that interventions provide the basis for scientific representations (Hacking 1983). Through the model's heterogeneous construction, representations and engagements were formed simultaneously and, moreover, jointly. Interaction between "technical" and "social" considerations fails to capture this relationship, in which causes are inseparable (see theme 5). Let me instead speak of scientists *representing-engaging*.

Theme 3. **Scientists are practically imaginative agents.** The idea of representing-engaging implies that scientific agents are mindful both of nature and of the social realms in which they act—in which they are situated— and that they project continuously between the situation studied and their social situation. In focusing on scientists' social *situatedness*, I am not saying that they are corrupt, fallible, lazy, or taking the path of least resistance. On the contrary, I am affirming that all human activity is imaginative—that is, the result of a labor process that grows out of the laborer's *imagination*. Agents assess, not necessarily explicitly, the practical constraints and facilitations of possible actions in advance of their acting (Robinson 1984).[16]

In fantasy, people envisage worlds and mentally inhabit them, escaping the practical difficulties of action. Imagination is not like that. Achieving some result in the material world requires human agents to be engaged with materials, tools, and other people. The KFM modeler had to engage with pasture growth, government sponsorship, an agricultural extension system, and so on. Moreover, materials, tools, and other people confront scientists with their recalcitrance. So scientists project themselves into possible engagements out in the world in order to imagine what will work easily for them and what will not. These constant projected confrontations with the components that personal and collective histories make available lie behind all the actions people take, including scientists' representing-engaging. Through their imagined engagements, people build up knowledge about their changing capabilities for acting in relation to the conditions in which they operate (though this knowledge may not be consciously articulated).

Theme 4. **The agency of heterogeneous constructors is distributed.** One consequence of focusing on agents' contingent and ongoing mobilizing of webs of materials, tools, and other people is that the character of their agency can be interpreted as *distributed* over those webs, not *concentrated* mentally inside socially autonomous units whose ideas or beliefs are key to the order they impart on the world. That is, although agents work with mental representations of their worlds and can speak about motivations, the malleability of those representations and motivations is not a matter of simply changing beliefs or rationality. Instead, a heterogeneous web of materials, tools, and other people help agents *act as if* the world were like their representations of it. During the Kerang study, the principal investigator may well have *believed* deeply that economic decision making was of primary importance in people's lives. However, he was able to sustain this belief against possible challenges through many *practical measures* (as discussed under themes 1–3). For example, although he knew about the sociological study of how farmers change, he did not secure access to it, and he concentrated on econometric investigations rather than developing skills in multi-objective analysis.

Theme 5. **Resources are causes.** Up to this point in my description of how the KFM was constructed, I have used the neutral term *component*. But there may be little explanatory significance to some of the diverse things that scientists link into webs as they establish support for their theories and ongoing scientific activity. Let me apply the term *resource* to components that make a claim or a course of action more difficult for others to modify. By extension, a resource for one person is a *constraint* for another person trying to modify the first's claim or action. Resources make a difference; that is, when resources are deployed, they function as *causes*. In this light, any descriptive use of the term *resource* also implies a claim about causes, and such claims invite analysis.[17]

Theme 6. **Counterfactuals are valuable for exposing causes.** The components of the construction process that I have chosen to mention were significant resources in the building of the KFM—or so my account of the KFM would imply. But how can I support the causal claims that I have thus structured into my account of the KFM? For a start, let me note that to support the causal claim that something made a difference logically requires an idea of what else could have been if the resource in question had been absent. That is, causal claims involve consideration of counterfactuals— things that might have occurred, but did not.

The sources for ideas about what else could have been are varied. Sociologists and historians of science undertake conceptual analyses or historical and cross-cultural comparisons (Harwood 2004) and give rein to their sociological imagination (Hughes 1971). They also listen to opposing parties in controversies (Collins 1981a, 1981b) and campaigns for social change (Nelkin 1984). Indeed, controversies and campaigns provide the clearest, most concrete evidence of alternatives because the agents themselves identify the resources they consider important.

There is no logical reason, however, why the resources explicitly exposed during a controversy should constitute the full set used by a scientist. There are resources taken for granted and shared by opposing parties, and, moreover, resources that must be mobilized even when there is no apparent controversy. In short, ideas of what else could have been should not be limited by whether anyone actually attempted to construct the alternative situation. For all these reasons, explicit use of counterfactuals may be needed in order to analyze a more inclusive array of resources used in the construction of science.

I began my account of the building of the KFM as a fairly neutral description. Notice, however, that I began introducing counterfactuals once I started to draw connections among the heterogeneous components. For example, in contrast to the single objective of maximizing income in the modeled farms, I mentioned the counterfactual possibility of multi-objective techniques. In explaining why such techniques were not incorporated into the KFM, I mentioned the principal investigator's training, his status relative to the modeler, the Institute's specialization, and the availability of software. These were constraints for anyone who might want to construct a multi-objective model. By identifying them, I was implying that the principal investigator's training and so on were resources for constructing a model with a single objective function. In this general fashion, exploring the practical constraints on *counterfactual* possibilities—what did not happen—can, by a logic of inversion, expose the resources that helped those who constructed what did happen (fig. 4.4). My analysis of the modeling of nomadic pastoralists in section B follows this logic (see chapter 5, note 20 for further discussion of the use of counterfactuals).

The emphasis on multiple, heterogeneous resources means that the relevant counterfactuals are multiple and particular. In principle, we could formulate some all-encompassing counterfactual. For example, an alternative to the Kerang study would be a project that was not conducive to

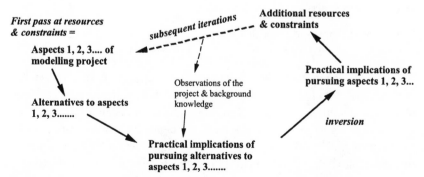

Figure 4.4 The method of exposing resources used in heterogeneous construction by exploring the practical constraints on counterfactual possibilities.

top-down government policymaking. However, if we were to consider the practical implications of such a counterfactual, we would be challenged to identify specific sites for possible modification of the research. This would be all the more the case if we focused on its practical implications for the specific scientific agents involved. In the case of the Kerang study, the modeler wanted to consider sociologically realistic processes of how farmers change. But his ability to produce results that paid attention to such processes was constrained by his distance—geographically, organizationally, and conceptually—from the farmers' domain of social action. This geographic and organizational distance was, in turn, related to the centralized character of government and intellectual activities in the one major city of each Australian state, something constrained by the previous 200 years of development. Toward the end of the project, the modeler contemplated a move counter to that centralization; namely, to live and work in the Kerang region as an agricultural consultant. He was aware that this move would raise practical issues such as purchase and maintenance of a car, long-distance access to computer facilities and libraries, ways to keep abreast of discussions about the wider state of the rural economy, and other considerations of a more personal nature. The modeler decided not to move, which meant that the representation of the Kerang region he was able to produce facilitated the making of policy based on simple economic grounds. This outcome did not flow from a political or intellectual commitment to economically based technocratic rationality; many practical, not just intellectual or ideological, considerations would have been entailed in producing a different result.

B. An Intersection of Domains of Action that Include MIT, USAID, System Dynamics Modelers, and Nomadic Pastoralists

How does technocratic rationality shape results when many practical considerations could have led the research in other directions?

In chapter 3, I noted a broad association between technocratic rationality and systems thinking. This picture is complicated by the view that scientists' work depends on their heterogeneous webs of resources. For technocratic rationality to result in technocratic management or governance, webs of practical considerations must come into play. Section B of this chapter, by exploring what it would mean in practice to pursue alternatives to systems thinking, seeks insight into a particular scientific construction of technocratic rationality.

Bringing Nomadic Pastoralism into the Domain of Modeling

By 1973, the semiarid Sahel region of West Africa had experienced five years of drought and developing crisis. Many pastoralists (livestock herders) and farmers were in refugee camps, their herds decimated and their crops having failed once again. Scenes of famine reached the European and American media, belatedly bringing the situation to international attention (Morentz 1980). Relief efforts were stepped up, and despite serious shortcomings in coordination and food distribution (Sheets and Morris 1974, 56; reprinted in Glantz 1976), massive starvation was averted. The United States, through its Agency for International Development (USAID), became the largest contributor to the relief effort.

Western commentators at the time focused not only on famine relief, but also on the causes of the crisis and on prospects for the region's future (USAID 1973; Dalby and Harrison Church 1973; Glantz 1976). Although conventional wisdom was that the postindependence governments lacked the capacity to cope with the situation, most researchers' assessments emphasized the natural conditions. The environment appeared fragile; the pastoralists and farmers had exceeded the carrying capacity of their land. The overgrazing of their herds and the cultivation of unsuitable lands had, in conjunction with the global cooling that climatologists detected (Bryson 1973; Lamb 1973; Winstanley 1973), set the desert marching southward. Some analysts saw the Sahelian drought and famine as a forerunner of further widespread population-resource crises to come; almost all agreed that the ecological resource base of the Sahel region had been seriously damaged. Once emergency relief was under way, discussion turned to longer-term measures to ensure recovery and prevent future disasters.

In this context, the United Nations (U.N.) convened a meeting in June 1973 on a mid- and long-term program for the region (United Nations 1973). Speaking on behalf of the United States, USAID administrator Donald Brown expressed willingness to commit resources beyond emergency requirements and proposed that "an American scientific institution begin preparation of . . . an initial analysis . . . of the technical problems and the major development possibilities" (Brown 1973). This proposal, accepted by the U.N. meeting, was endorsed the following month as part of a $30 million congressional authorization for relief and rehabilitation assistance in Sahelian Africa (U.S. Senate 1973). USAID then funded a one-year, $1 million project at the Massachusetts Institute of Technology (MIT) to evaluate long-term development strategies for the Sahel and the bordering "Sudan" region (Seifert and Kamrany 1974).

One component of this project was a study of nomadic pastoralists. These livestock herders spend part of each year moving with their livestock over the range in search of pasture. Such migration is necessary because rainfall in the Sahel is patchy in distribution and varies greatly from year to year, dramatically affecting the location of good pasture. After a three-week visit to the region, a graduate student at MIT, Tony Picardi, with a background in systems analyses of population and ecological issues, constructed and reported on a sequence of system dynamics models "for understanding the ecological and social dynamics of the pastoral system" (Picardi 1974c, abstract, i). Picardi's models of pastoralists included many factors and mathematical relationships (a schema of the simplest model is given in fig. 4.5). Yet he summarized his findings simply, in terms of the "tragedy of the commons" (Hardin 1968): Each herder with access to common rangeland follows the same logic: "I will receive the benefit in the short run from increasing my herd by one animal; everyone will share any cost of diminished pasture per animal; therefore I will add another animal to my herd." Overstocking and overgrazing was thus inevitable. Soil degradation and eventual desertification could be avoided only if all the pastoralists replaced their individual self-interest with "long-term preservation of the resource base as their first priority" (Picardi and Seifert 1976, 51), perhaps entering ranching schemes that privatized or strictly supervised access to pasture (Picardi 1974c, 55–59, 165–68). (For more recent assessments of common resources and privatization, see chapter 6.)

As other researchers have come to emphasize, nomadic pastoralism is not one undifferentiated phenomenon. Pastoralists have to cope with variable and uncertain semiarid environments. To different degrees in different locations, they also have to cope with an extension of the area under crops, privatization of access to resources, government policies that usurp traditional systems of authority and conflict resolution, and development projects that favor sedentary over nomadic ways of life (Horowitz and Little 1987). Pastoralists make their living within such diverse environmental and social processes, working directly or indirectly with agents at the local to the international levels. These interactions and the pastoralists' knowledge of them provide resources, which are diverse and changing over time. In short, the pastoralists also build heterogeneous webs.

Now, in making science of the pastoralists' practices, Picardi had to ensure that the heterogeneity and change of these practices did not give someone grounds for disputing his models as representations of reality. When one asks who "someone" might be, Picardi's task appears even larger. Intersecting in Picardi's modeling were many agents who otherwise

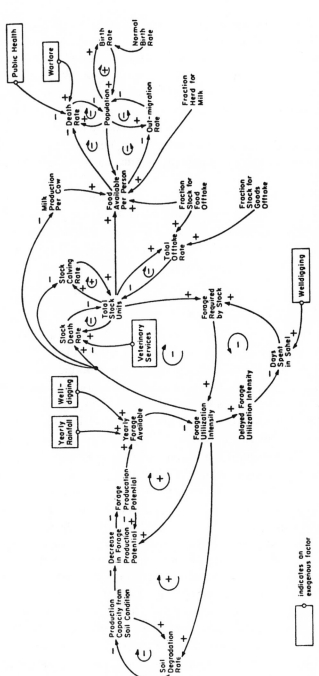

Figure 4.5 A schema of SAHEL2—a model of pasture, livestock, and pastoralist populations. Changes in variables are governed by equations incorporating effects from all the variables linked to them. The signs on the arrows denote the effect of a link; the signs inside curved arrows denote the overall effect of the links in a feedback loop. (From Picardi 1974c.)

operated in diverse domains of social action: the various researchers at MIT, USAID, the U.S. Congress, Africanists, African officials, and the pastoralists themselves. Picardi's models would be promoted or discounted according to whether or not they could become resources for these other agents' own actions. In other words, when Picardi was modeling, he was not only representing, but also engaging within these intersecting domains of action. To ensure support for his actions, he also had to stabilize, at least for some time, the diversity of agents who, through their own representing and engaging, now intersected, but might later diverge from, Picardi's making of science.

I use the counterfactual method introduced in section A to reconstruct Picardi's modeling work. This method enables me to expose the web of resources that he mobilized to produce his scientific representations of the ecology and society of nomadic pastoralists. First, however, let me provide some more background to Picardi's work.

Sketching a Heterogeneous Web

More than a year before the Sahel captured international attention, the USAID deputy administrator for Africa, Samuel Adams, initiated an inquiry into measures beyond food aid and changes in food production and marketing aimed at reversing the deterioration of the region's resource base. Could the region be lifted out of its marginal and declining status and transformed by U.S. assistance based on infusions of capital and modern science, such as remote sensing by satellite, new plant varieties, and intensive livestock production systems? The famine of 1973 brought congressional support to USAID's efforts, and the U.N. meeting provided international endorsement for a study of long-term development alternatives. USAID recognized that the report would have to be excellent because some reservations had been expressed at the U.N. meeting about whether the U.S. proposal could produce useful results. Moreover, a short time frame had been set for subsequent meetings of international donors and African governments, so the study would have to proceed rapidly. MIT was given a year and $1 million to produce the report for USAID (Lyman 1976).

USAID had clear reasons for selecting MIT as the scientific institution to conduct the analysis of long-term development strategies. As USAID administrator M. J. Harvey explained in a letter to a congressman (written probably by Princeton Lyman, then director of USAID's Office of Development Services and the USAID official most closely involved with establishing and evaluating the project; see Lyman 1974, 1976):

[MIT] had experience in systems approaches, especially with regard to water systems. It alone, of the institutions investigated, offered to devote the attention of senior people over long periods to the task, and it had the stature needed to attract French, African, and other U.S. academic cooperation in the task as well as to substantiate the validity of such an approach. (Harvey 1974)

Although he had no previous experience in Africa, MIT appointed William Seifert, an experienced systems analyst and administrator, as the principal investigator of the project. The project was to be jointly administered by MIT's Department of Civil Engineering, in which Seifert was a professor, and the Center for Policy Alternatives. The Center had only recently been established to undertake interdisciplinary policy studies, and this was its first major project. Seifert's technical and administrative background was in control systems, particularly for weapons, but he had turned his attention to the management of complex systems more generally, including, in the early 1960s, transport systems and then resource management. He had initiated an innovative "systems engineering" seminar, which entailed students forming teams to undertake case studies of problems having not only technical but also social, economic, and political dimensions, such as Project Bosporus on future airport and seaport developments for Boston (Seifert and Corones 1970). These projects, while not commissioned by actual clients, were semiprofessional; the reports were published by the MIT Press after presentation at "conferences" with guests from business and government.

System was a central but equivocal term. A system could minimally denote an orderly collection of interacting components, and a systems approach, a juxtaposition of different perspectives on the same system (Seifert and Corones 1970). Systems analysts usually qualified this minimal definition by referring to the problem or the purpose of the study or to their clients; that is, systems analysis not only represented the nature of the system, but did so in relation to someone's conception of possible interventions. At the same time that Seifert promoted a very broad or "loose" definition of the systems approach, he also advocated "tight" methods— modeling and other quantitative techniques—as means of deriving at least some of the different perspectives. USAID's concept of a systems approach, which the head of the Center for Policy Alternatives had helped USAID shape, included both levels. At the loose level, MIT was to convene a retreat of experts who would collaborate to expose new aspects and linkages of the situation and establish the initial framework for the project. In fact, USAID rationalized that MIT's lack of African specialists might be an asset at this stage, for it might help these experts avoid preconceptions of the situation. After the experts narrowed and sharpened the development alternatives, more specialized and sophisticated methods of systems analysis would be brought into play (Lyman 1976).

Regardless of the open-mindedness implied by the loose systems approach, *system* appeared to have stronger connotations for Seifert and other systems engineers at MIT. Let me try to make these strong connotations explicit. (The quoted terms to follow were common to many authors, including Seifert and Picardi; see Picardi 1974c and Seifert and Kamrany 1974.) A system was an enduring entity, an object susceptible not only to systematic investigation but also, given some problem or goal, to management or even to engineering. Both living and nonliving, conscious and mechanical, could be combined in one systems analysis. A system had "dynamics" or "behavior" produced by interactions or "feedback loops." It had to be studied as a whole because its dynamics were inaccessible by "illegitimate isolation" of variables; that is, by examining individual components and their "simple cause-and-effect" interactions. The "forces" acting in a system were larger than the individual. Moreover, the particularity of individuals tended to obscure the "generic" behavior and "underlying mechanics" common to many systems. This holist, supra-individual framework, together with the mechanical or behaviorist terminology, suggested a special role for analysts able to observe the system as a whole from the outside, maintaining an appropriate scientific distance or managerial overview (chapter 3).

Seifert introduced Picardi to systems analyses and sponsored his program of interdisciplinary studies, with topics ranging from gypsy moth outbreaks through the management of family planning programs among the poor to the economics of a zero-growth society. Seifert involved Picardi in a UNESCO-funded study to model developing countries. Picardi's master's thesis, a demographic and economic model of Bolivia, emerged from this study. He noted that Bolivia was a suitable system for studying "interactions between population, economics and natural resources" because it had "relatively few complex ties with the outside world" (Picardi 1973). While Picardi waited for Seifert to secure funding for another system study that could cover a doctoral dissertation, he had to support himself by working for other professors on models of heated water discharge in estuaries and movement of toxic chemicals through the environment. But as soon as Seifert was granted the Sahel-Sudan contract, Picardi was brought in and began to carve out a relevant project for his dissertation research.

Although Seifert advocated computer modeling of systems, he was not a practitioner. Picardi had learned his modeling methodology, system dynamics (SD), from its originator, Jay Forrester, and from others working with or trained by Forrester. Like Seifert, Forrester had moved from engineering and defense-related work in the 1940s and 1950s into management and, in the late 1960s, into population and resource issues. He developed SD in the context of improving the management of firms that experienced cycles of surplus inventory alternating with backlogs of unmet orders. Forrester extended SD to capture the dynamics of urban growth and decay and, in the late 1960s, built a global model that attracted funding from the Club of Rome, an elite group of Western government, business, and scientific leaders. The report that emerged, published as *The Limits to Growth* (Meadows et al. 1972), predicted global population and economic collapse unless universal no-growth policies were immediately established. This study made SD famous: Forrester's System Dynamics Group was the most active center of systems research at MIT in the early 1970s (Bloomfield 1986).

Picardi picked up many methods for analyzing systems, including input/output analysis, econometrics, operations research, action-impact matrices, and public expenditure theory, but he found SD the most exciting. As he recalls it (Picardi 1989a), SD allowed a modeler to incorporate "unquantifiable social factors," "thought processes," and "people's perceptions." Good system dynamicists were those who got beyond the "physical stuff" because the "dynamics of systems, when you have people in them, are ones of values and long-term beliefs interacting with the physical environment." Yet, although Picardi used SD and became proficient with its computer implementation, DYNAMO (Pugh 1973), he did not join the System Dynamics Group. He found "Forrester's crowd" dogmatic about the use of SD and not inclined to base their systems analyses on detailed, historically derived data (Picardi 1989a).

Seifert and Forrester were by no means the only peaks in the landscape of systems analysis at MIT. Picardi also attended, for example, the seminars of Carroll Wilson on sustainable growth and the steady state society. Wilson, postwar head of the Atomic Energy Commission, had later directed studies on international security and the application of science and technology in developing countries. And, in 1970, he organized a month-long gathering of international experts to study critical global environmental problems and human effects on climate (Study of Critical Environmental Problems 1970). Wilson's international, interdisciplinary projects had provided a model that the MIT officials could present to USAID to help secure the Sahel-Sudan contract.

MIT, in short, had senior personnel with experience in organizing multi-disciplinary, international studies, men who confidently used a language that fused science and management in analyzing multifaceted "systems." It could readily appear to be the right site for USAID's job. As Picardi (1989a) recalls it, "MIT was the place you came to solve problems."

USAID, however, did not leave the job entirely in MIT's hands. It organized briefings in Washington for the research team on Africa and recruited Africans and French Africanists to come to work in the United States. Eventually, USAID's mediating role would presumably have to work in the other direction, transmitting the results of the study to Congress, to international bodies, and, through its own development projects, to Africans. In the meantime, USAID assumed an advisory and reviewing position. After setting the terms of the study, it signed the contract and left the specifics of the systems analysis to the MIT study team.

Probing a Heterogeneous Web

The preceding narrative could be continued. I could connect Picardi's definition of the problem he had to solve, the briefings in Washington, contacts made during his three weeks in Paris and West Africa, collaborations and disagreements back at MIT, the sequence of models developed, and so on, to weave a rich picture of Picardi's modeling work. In a style similar to that of chapter 3's account of H. T. Odum's ecology, I might tell a plausible story to the effect that the models resulting from Picardi's work were appropriate for the context in which they were produced. Projected into the models was the theme that pastoralists, like mortals in a Greek tragedy, unwittingly acted to bring about their own ruin. They needed guidance from an outside agency—specifically, USAID, in turn guided by outsiders at MIT capable of understanding the system as a whole. Moreover, Picardi's models demonstrated from many different angles that the crisis was severe and could not readily be averted; this conclusion secured a voice for the young researcher, whose work might otherwise have been marginal to the future of the pastoralists.

At this point, however, I want to interrupt the narrative-making and tease out the different contextual connections facilitating Picardi's work. Let me explain this shift. Narrative presentations such as the preceding section are never mere descriptions. By selection and juxtaposition, they assign weight to different factors and make implications about causality. This is evident, for example, when I point to the connotations of "system" as an entity subject to both systematic investigation and management. I am not merely describing Picardi's context, but implying that these connotations of "system" might be a resource that enabled Picardi—an outsider and new-comer—to imagine he could gain insights unavailable to the pastoralists or perhaps even to Africanists (Picardi 1974a). The language of systems becomes one condition that makes a difference and which provides, not alone but together with other conditions, a sufficient set to account for this episode of modeling. Of course, despite the implications that narratives make about causality and about what else could have been, they are cast in terms of what did happen. Moreover, even when the course of events is a contested one, resources that are shared by the different parties are often taken for granted. I want, therefore, to analyze the specific implications about causality in the production of Picardi's models and attempt to iden-tify how the research could have been done differently.

My analysis is guided by the themes from section A of this chapter. Let me start with resources being causes, things that make a difference. Causal claims thus require ideas about what else could have been. Now, Picardi's models were not strongly contested at the time they were produced. Later, when they were (Brokensha, Horowitz, and Scudder 1977), Picardi was working on different issues from another region of the world. Therefore, in the absence of a controversy involving Picardi, I had to identify alternatives to the various aspects of system dynamics by some other means. Over a period of four years (1985–1989), I reviewed applications and critiques of system dynamics, attended classes and conducted interviews with the System Dynamics Group at MIT, manipulated Picardi's model on a com-puter, and reviewed subsequent analyses of nomadic pastoralists in sub-Saharan Africa. (See Turner 2003 for an entry point into subsequent research.) From these different sources, I distilled eight consistent aspects of a "strong" SD view of systems. Finally, I characterized alternatives to each aspect.

The contrasts in Table 4.1 between SD and the alternatives are not meant to be read as indicating that Picardi's models were limited techni-cally. One could speculate that a better modeler could have overcome these limitations, if not in 1974, then sometime subsequently when further technical advances had been made. However, if scientists represent-engage, we need to examine the varied and interconnected actions taken by scientists. What would it have meant *practically* to pursue the alterna-tives, given the many domains of action intersecting in Picardi's making of SD models? By exploring the implications of pursuing these counterfactu-als, I am able to expose the ways in which Picardi's work was facilitated by his staying with the conventional aspects of SD.

Table 4.1. Eight consistent aspects of system dynamics modeling and contrasting alternatives.

Aspect of modeling	Character of system dynamics	Alternatives
1. Rules and system structure	Fixed	Changing
2. History as a source of	Long-term values	Conditions for future changes
3. Particularity	One generic system	Locally particular systems
4. Individuals	Uniform and aggregated	Stratified and differentiating
5. Parameters	Constant	Constructed
6. Temporal and spatial variability	Leaves system structure unchanged	Essential to system structure
7a. Systems	Partitionable into subsystems	Not necessarily partitionable
7b. External forces	Simply mediated	Contribute to "internal" restructuring
8. Responses to crises require	Overall policy changes	Local participation in responses

In identifying these interdependent features of SD, I built on my personal experience as an ecological and environmental modeler and as an interpreter of modeling. I cannot claim to be neutral; another analyst might identify different counterfactuals. (I return to this issue in chapter 5.) Moreover, the alternatives in Table 4.1 are not exhaustive; I claim simply that they are economically framed for the task of exposing and evaluating the resources Picardi built on in his heterogeneous web. Such analysis helps provide a fuller understanding of why Picardi modeled as he did.

Let me list the eight aspects and consider each in turn, illustrating the alternatives using abbreviated examples (given in boxes) from research on African pastoralism.

Contrast 1: Rules and system structure fixed versus changing. Picardi designed his models "for understanding the ecological and social dynamics of the pastoral system" (Picardi 1974c, i). For Picardi, as for other system dynamicists, it was unquestioned that the world—at whatever level of resolution he used—is composed of systems. In SD, a system connotes more than an orderly collection of interacting components subject to scientific management. A system in SD is a bounded integrated entity, the behavior of which is primarily determined by internal interactions or rules (Picardi 1974c, 4, 7, 19ff; Forrester 1969, 17ff). External factors are simply mediated—for example, when energy comes into an ecosystem or people migrate out of a pastoralist society (Picardi 1974c, 7, 15; see contrast 7 below).

In contrast to analyzing complex interactions as self-determining and enduring, one might analyze changes in the structure of those interactions and in the rules governing them. This restructuring can be illustrated with an example from research on pastoralism in another part of Africa.

During the mid-nineteenth century, the Fulbe peoples of what is now Mali codified conventions for land use and access in the floodplain of the inland Niger delta. This code, or Dina, divided the floodplain into clan lands. A jooro, or tax collector/pasture manager, for each clan controlled access for livestock from other clans, in particular the timing of access. Under the colonial and postcolonial governments, the jooro have less power to enforce their control over land use and access. Rice cultivators, for example, have encroached on lands traditionally grazed by the pastoralists' livestock. At the same time, more jooro have begun to extract monetarized taxes for their personal benefit, further reducing their authority to regulate land use (Turner 1993; see Little 1987 for a historical review of changes in East African pastoralism).

In principle, modelers could incorporate changing rules into their models. To do so, they would need to anticipate the restructuring that may result from crises—such as the loss of livestock during the 1968–1974 drought in the Sahel—or from external interventions—such as the administrative actions progressively undermining the Dina in Mali. The modeler would then incorporate the range of system structures into the model from the outset and specify transitions or switches among those structures. In practice, however, such prior specification is difficult. It is no trivial issue for pastoralists to anticipate the new arrangements they will make when, say, they rebuild their herds after a drought or react to encroachment. An outsider who wanted to anticipate structural change might live with pastoralists for long enough to observe how they respond to change or undertake a detailed comparative study to see how other pastoralists had responded to similar situations. Given the short study time dictated by USAID, Picardi did not follow either of these courses. He considered only a small number of switches within the model system, corresponding predominantly to policy changes such as initiation of taxation to enforce destocking (Picardi 1974c, 323ff). These changes were to be imposed from the outside, not generated by the pastoralists.

Although USAID's short study time constrained Picardi's modeling to discount restructuring, it also facilitated his work. The time limit meant that he could not be expected to undertake a more detailed study and develop a sustained engagement with pastoralists. Furthermore, USAID had requested an evaluation of long-term strategies for the region, intending to use the results to advise the U.S. Congress and the United Nations in assisting the region. Specific strategies for international intervention were called for. Picardi was well aware of the need to communicate his results to clients (Picardi 1974c, 4, 6, 19, 216–17). Moreover, if Picardi's evaluation of nomadic pastoralism had been replete with pathways branching according to possible restructuring of arrangements, then significant translation would have been expected of Picardi by the project's sponsors, USAID (or of USAID by its sponsors). Translation would especially be needed if the possible restructuring depended on future initiatives of the pastoralists.

Of course, Picardi's actions were not determined by this one relationship with USAID. As we shall see in the following and later contrasts, other aspects of work relations, tool use, and language were implicated in emphasizing system-ness and deemphasizing restructuring.

Contrast 2: History as a source of long-term values versus conditions for future changes. Picardi's simulations began with the year 1920 and included historical periods of low and high rainfall as well as estimated increases in usable rangeland, stock survival, and human life expectancy due, respectively, to well digging, veterinary services, and public health expenditure. Apart from these parameter changes, however, the model system extended unchanged back to 1920, just as it continued with minimal switches in structure into the future.

An alternative construction of history, which reinforces the previous issue of restructuring, could work to remind a modeler that relationships have changed and, by inference, may not persist as they appear now into the future. Moreover, such future changes might be born of multifaceted social changes set in motion in the past.

Frantz (1981) described pastoral Fulbe who a century ago colonized a higher-elevation montane grassland area straddling what is now the Nigeria-Cameroon border. By the 1940s, increased cattle and human populations led colonial officials to regulate the location and size of herds during the rainy season migrations. No longer could entire households move with their herds as far as necessary to find pasture. Instead, pastoralists began to subdivide their herds and arrange for pasturage outside their authorized area. Households became more sedentary, with hired herdsmen, increasingly non-kinsmen, or even non-Fulbe, overseeing the more restricted movements of the dispersed subsections of the herds. Dry season herd locations have also become regulated and limited by agricultural expansion, resulting in longer herd migrations than before. In recent decades, the pastoral Fulbe have begun to supply meat for consumption in distant urban centers, further widening the spheres of social interaction into which they are linked and the range of conditions contributing to their future responses.

Detailed historical research is needed before such complexity of social arrangements and rearrangements can be incorporated into any model or other representation. Not only was detailed information on pastoralism scarce in 1974 (Swift 1977), but Picardi's study time was limited. Again, these "constraints" facilitated Picardi's modeling. By omitting such social complexity, Picardi could reduce history to a set of distinct perturbations to a system. In this sense, the rangeland was a system in which demographic pressure builds up as population approaches the carrying capacity, which is set by the area and rainfall (Picardi 1974c, 45–60). Warfare, Picardi remarked, "represented a significant negative demographic pressure" (46) until curtailed by colonial rule. Given that "social values" or "norms" (137–65) were fixed or changed too slowly to circumvent the Malthusian equation (2), pastoralism's history and future seemed determined by the environment. In this spirit, Picardi interpreted a green area detected in satellite photos as the result of "a fence and simple management policies" (57) without delving into the history of how the area—a French-run cattle and sheep ranch (Sheets and Morris 1974, 67)—was alienated from the pastoralists' use. For the MIT researchers and their sponsors, working several thousand kilometers from the region and unable or unwilling to become immersed in West African societies, environmental determination facilitated the formulation of general analyses and policies.

Closer to home, Picardi, a self-assured but junior member of the team, had to negotiate a relationship with his more senior colleague, Len

Cockrum, the range biologist on the study team and a member of Picardi's dissertation committee. Cockrum believed that scientific study should be restricted to the pasture and livestock, and he disapproved of including people as variables in a pastoralism model. Incorporating more social complexity into the models than Picardi did would have generated even stronger disagreement from Cockrum. (As it was, Picardi eventually had to enlist Seifert's assistance to finagle Cockrum's acceptance of the dissertation because Picardi had introduced some equations relating to "social and economic motivations"; Picardi 1974c, 44; 1989a.)

On the other hand, anthropologists, sociologists, Africanists, or visiting Africans might have sensitized Picardi to the complexity of the pastoralists' social interactions. The loose organization of the research militated against this possibility. Under Seifert's direction, the project members worked quite independently. The retreat of experts that USAID expected was never convened, and no one established a framework for interdisciplinary collaboration among the researchers. The project's anthropologist was preoccupied with completing some previous project. A sociologist spent three weeks at MIT, arriving ten months after the project began. Neither responded to Picardi's memorandum seeking answers to his questions regarding pastoralists' range management practices (Picardi 1974a, 1974b). Both USAID and MIT failed to recruit Africans to work in the United States. In the absence of anyone insisting on the complexity of pastoralist societies, Picardi considered it "appropriate to use an ecologically oriented model," interpreting his mission as analyzing "policies that directly affect the ecological and livestock systems" (Picardi 1974c, 31). The behavior of humans in Picardi's models was effectively determined by their environment, which brought Picardi and Cockrum closer in line than their expressed disagreement suggested.

Contrast 3: Generic systems versus local particularity. Picardi stated his aim as modeling "the pastoral system." The definite article here signals his assumption that underlying nomadic pastoralism was one phenomenon, everywhere consisting of the same fundamental processes. Differences from place to place were discounted. Indeed, Picardi claimed that a "model must be highly aggregated . . . to be useful for understanding and communicating fundamental processes [and should focus] on generic behavior patterns" (Picardi 1974c, 4). In contrast, a modeler might identify a diversity of pastoral situations and consider how their particularities influence their development.

The pastoral Fulbe in other areas of Nigeria and Cameroon share with the montane Fulbe in the previous example a progressive loss of access to grazing land because of agricultural expansion. They differ from the montane Fulbe, however, in being influenced by mining, industry, the construction of dams, and more frequent droughts. Their herds have to migrate farther and into more humid zones where disease incidence is higher. Their economy is also more strongly intertwined with that of farmers and townsmen who, far more than in the montane areas, exchange crop products, residues, and other commodities with the pastoralists for livestock products (Frantz 1981; see also Galaty and Johnson 1990 for a sense of the historical and geographic variation among nomadic pastoralists).

Some of the practical implications of modeling a diversity of pastoral situations have already been mentioned. Accommodating this alternative would require, as is the case for restructuring and history (contrasts 1 and 2), more time, translation from particular analyses to general policy, and greater involvement by anthropologists, Africanists, and so on. Conversely, modeling pastoralism as one system allowed Picardi to fulfill the sponsor's requirements for clearly characterizable strategies of intervention. It also allowed him to accommodate the minimal collaboration of the project's social scientists. In addition, a modeler of local particularity would have to assemble detailed information from a diversity of localities. After one initial visit to the region, USAID did not fund further African trips by the MIT researchers. Picardi was fortunate to receive from French researchers a detailed census of one locality in western Niger. With the assumption that a generic model could be built, this one locality became not merely a particular place, but a source of data on which to base a model of pastoralism in general.

Contrast 4: Uniform and aggregated individuals versus strata of differentiating individuals. In Picardi's models, all individuals—households, livestock, or plants—behaved identically; the prospects for pastoralists in the region were a simple multiple of the output of the model. In contrast, a modeler could consider the effects of differences in the wealth and power of households.

In principle, it is simple for a modeler to distinguish different strata of individuals and specify their characteristics. More detailed data would, of

Little (1985, 1987) describes the differentiation of Il Chamus pastoralists in an area of Kenya whose ecology is similar to the West African Sahel. Having suffered through prolonged droughts during the 1970s and into the 1980s, poor herders engaged in risky, but inexpensive, dryland (rainfed) farming in order to survive. Wealthy herders subject to the same drought could afford the labor and capital to engage in irrigated agriculture and thus reduce the need to sell livestock for grain during dry periods. After a drought, the rich herder-agriculturalists could rebuild their herds more rapidly; some of the poor became their hired laborers. The differentiation among pastoralists has been accentuated by the fact that rich herders command greater influence in land claims once states initiate privatization of land holdings. Now that there has been an increase in cultivation and wage-earning activities, labor for herding has become a limiting factor. Rich herders can pay for their herds to be grazed on better land some distance away from settlements, while the poor, who must make use of wage-earning opportunities, graze their herds near their households. As a result, environmental degradation, where apparent, lies close to population concentrations—not, contrary to the tragedy of the commons view of nomadic pastoralism, out on the range.

course, be needed. In addition, an alternative to SD—or patience in using it—would be required because SD's computer implementation, DYNAMO, was not designed for multiple variants (arrays) of each basic variable. The speed of computer operations and the clarity of the diagrams used to illustrate the system drops rapidly as variables proliferate. (These characteristics persist in STELLA, the most popular current software for SD.) On the other hand, while modeling strata is conceptually simple, modeling the process of differentiation of strata remains more difficult. The characteristics of the strata change as they accumulate or become impoverished. In fact, the very structure of the system may change (e.g., herders becoming agriculturalists and wage laborers). As in the case of restructuring (contrast 1), the modeler would have to anticipate these changes.

To model differentiation, Picardi would also have had to work without exemplars to follow because in 1974 there were no SD models of differentiation. (The situation is not so different today, but see Bhaduri 1983; Saeed 1982.) Nor were data pertaining to differentiation in West Africa available (see Sutter 1983). The uniform model of pastoralists Picardi used obviated any need for data on strata or differentiation, thus facilitating his work in the same manner as omitting local particularity (contrast 3). Stratification was, in any case, less apparent in the locality Picardi chose as the basis of his model. In that locality, Tuareg pastoralists had not become sedentary, nor were they deeply implicated in the agricultural economy. They were thus closer to the pure or generic system desired by systems analysts.

Uniformity of model individuals facilitated Picardi's modeling in an additional way. When system dynamicists seek to establish the realism of their models, their prime means of persuasion is not to demonstrate close correspondence of model predictions with actual observations. Instead, they render their models plausible by directing their presentations at non-specialists in the area in question and drawing them into the logic of the model. The rationality of modeled individuals is validated by the listener's personal experience—Would you decide any differently in the same circumstances? (Picardi 1974c, 199). The system dynamicist then uses SD—often reinforced through computer games (Sterman 1987)—to demonstrate that locally rational decisions, when worked through feedbacks in the models, generate unanticipated and counterproductive outcomes. The scenario of Hardin's "tragedy of the commons," which was represented for the first time in SD terms by Picardi's model (Picardi 1974c, abstract, i), has achieved widespread recognition by the same means (see chapter 6, section A).

If a SD model were to specify heterogeneous individuals, its realism would be harder to establish by personal validation and weight of logic. The outcomes would no longer be simple and inexorable—which of the strata would the listener identify with? With all pastoralists alike, persuasion by logic was possible. Direct empirical evidence of selfish individual exploitation of the common range was not needed (Picardi 1974c, 162ff). Contrary or more complex possibilities proposed by specialists could remain out of Picardi's picture—for example, labor demands rather than range area limiting expansion of herds (Brokensha, Horowitz, and Scudder 1977); accumulation and impoverishment driving environmental degradation (Little 1985; see contrast 3 and chapter 6, section A). USAID, in turn, was spared the difficult, politically charged task of considering explicitly how its programs of assistance and support for state policies would differentially affect pastoralists who had unequal access to resources—in particular, to the "common" rangeland.

Contrast 5: Constant versus constructed parameters. For Picardi, history was a record of "long-term social values" and of a "traditional economic system" (Picardi 1974c, abstract, ii). The fertility of women, the fraction of livestock sold each year, and so on were effectively constant parameters in the model (70–71). The fertility parameter could change only slowly as life expectancy was perceived to increase; an enforceable taxation policy was required to increase livestock offtake. In contrast, parameters might be more susceptible to change.

> Swift (1977) concludes a review of pastoralist demographics by noting that many factors contribute to the low natural rates of population increase for nomadic pastoralists, including high male-to-female ratios, late marriages, men away herding or on labor migrations, and long breast feeding of infants.

In the example above, the birth rate could move up or down relatively rapidly following sedentarization of pastoralists or transition from herding to other kinds of labor that required more frequent absences of men from the household. Fertility could therefore be thought of as a variable constructed of other parameters—a variable whose future constancy would not be guaranteed by its historical invariance.

A modeler could accommodate this constructedness of parameters either by enriching the model's structure so that underlying factors were included or by revising the model whenever previously constant parameters revealed their dependence on underlying determinants. The first modification would undermine the mechanical determinism of the typical system model. The second would shorten the time horizon of confident projections and require ongoing assessment of the modeled situation. On the other hand, Picardi was able, by limiting the deconstruction of parameters, to preserve the strong SD view of system with its special role for the outside analyst. The models could be used by researchers based in the United States to make projections for evaluating the long-term prospects of the region, as required by USAID. By not requiring ongoing assessment, Picardi could complete the evaluation in the fixed period established by USAID.

Contrast 6: Temporal and spatial variability leaves system structure unchanged versus essential to system structure. The variability over time and space of rainfall and pasture production in the Sahel was well recognized in the early 1970s. In a uniform model, however, all pastoralists must experience this variability and uncertainty identically, as external factors driving the model. In contrast, this variability could enter into the very structure of a model.

For the Il Chamus described by Little (1985), droughts increase the demand for grain, and any cattle fit to be sold fetch lower prices. Rich herders have sufficient grain in storage to keep their stock off the market, while poor herders have to buy grain and sell their livestock at just that time when the terms of trade have turned against them. Given that cattle prices rise when the drought breaks, poor herders are further squeezed and may turn to wage labor, further constraining their herding. Little (1985) reports such a dynamic even during a normal year's preharvest seasons.

The practical implications of modeling variability are similar to those of modeling the heterogeneity of individuals (contrast 4)—more data and computational power are needed, and it is difficult to persuade listeners by leading them through the logic of the model. In addition, even if differentiation is overlooked, spatial and temporal variability require decisions by the modeler about scale. Exchanges and transfers of goods and services among different regions and across seasons or years can be included only if the model is subdivided into regions with different relations specified for the different seasons, and so on. At higher levels of averaging or aggregation, however, these transfers are hidden. In general, when systems models are resolved at finer scales, they usually exhibit more complex mathematical behavior. It becomes more difficult to discern generic modes, such as the ubiquitous pattern in SD of exponential growth followed by a crash and cycles of recovery and decline. Without intertemporal and interspatial transfers, however, Picardi was spared not only the more complex mathematical analysis, but also the duty to translate such analysis so that USAID could comprehend it.

Contrast 7: Partitionable systems in which external forces have simply mediated effects across the boundaries versus nonpartitionable systems in which "external" forces make for "internal" restructuring. Picardi's "working hypothesis" was "that the problem behavior of the ecological-pastoral system (the desertification and recurring famines) results primarily from processes at work within the system" (Picardi 1974c, 7). For Picardi, pastoralism was an instance of the world being partitionable into systems and systems within systems. The inside and outside of the system in question could be well demarcated—Picardi chose a locality in which nomadic pastoralism appeared to be practiced in pure form, in that herders were not also farmers (see contrast 4)—and the dynamics of the system were effectively unlinked from the dynamics of its context or layers of context. Although the pastoral system was not closed, the system's relationship with external factors was simply mediated—pastoralists were able to emigrate, but not to change, say, into pastoralist-laborers. Policy changes initiated outside the system (e.g., veterinary services) had accentuated the "processes at work within the system." To overcome the problems of desertification and famine, policy based on an understanding of the system as a whole was needed.

A modeler could find it more difficult to divide a situation into inside and outside. Most pastoralists are engaged with other groups—at the very least, with agriculturalists with whom they trade for grain and arrange for access to crop stubble, on which the herds graze and in return deposit manure. Some pastoralists are, moreover, pastoralist-farmers or farmer-pastoralists (Little 1987). In fact, the very existence of any relatively self-contained or "pure" instance of nomadic pastoralism could be taken as a special outcome to be explained—for example, as a result of pastoralists moving to avoid raiding or taxation—rather than as a starting point for theory. The anthropologist Eric Wolf recommends visualizing social forms as "historically changing, imperfectly bounded, multiple branching social alignments" whose explanation should take into account the wider field of economic and political forces (Wolf 1982, 387).

A modeler, therefore, might analyze the potential for external forces to restructure internal relations (see contrast 1) and treat the boundary between outside and inside processes as problematic. For example, the marginalization of poor herders (see contrast 6) has been occurring not only as a result of adverse terms of trade during a drought or seasonal cycle—the relative prices of grain, meat, and so on—but also as a result of a longer-term deterioration of terms of trade for the Il Chamus (Little 1985). The future prospects of these pastoralists depend on this trend, and its explanation would encompass factors beyond the boundaries of the pastoral system, as far even as the international terms of trade.

Nonpartitionability threatens the very project of socio-environmental modeling. If a system cannot be assumed to be separable from its context, the modelers have to appraise whether the context confounds the interactions specified in the model, even if they have elevated the system only temporarily out of its context. Furthermore, the historical contingency of changes in context calls for the modelers to continually assess the situation, rather than making confident projections at any one time. Although no quantitative modeling methodology existed in 1974 to address the challenge of context sensitivity, I might note that methods developed since require a team of researchers with a variety of skills (Walters 1986). Picardi, however, was under instruction to carve out a doctoral project to be completed in one year. He was working as a junior researcher within a group that Seifert had not organized into a coherent team. Accepting that systems can be partitioned into subsystems helped him to complete his part of the project without having to renegotiate either his task or the project's internal organization, and without having to develop a new vocabulary suited to the non-systemness and context sensitivity of socio-environmental situations.

Even if such a modeling methodology were available, modelers who did not clearly divide the external from the internal would face the same practical considerations mentioned for restructuring and the inclusion of historically contingent social complexity (contrasts 1 and 2). On a more subtle level, analysts would not so readily be able to locate themselves outside the system. They would become part of the context and a potential influence on the system. For Picardi, however, it was not a problem to assume an external position. Doing so allowed him to observe scientifically and evaluate policy objectively without needing to examine his own interests in the situation, its representation, and its management. USAID, in turn, received the combination of science and policy assessment it sought (Lyman 1976).

Contrast 8: Responses to crises require overall policy changes versus local participation. In Picardi's analysis, interventions initiated by outsiders, such as veterinary and public health efforts, pacification, and well digging (Picardi 1974c, 49ff), together with the high-rainfall period of the late 1950s and early 1960s, had accelerated the increase in livestock numbers and primed the system for the crash that occurred during the 1968–1974 drought. Insiders—the pastoralists—provided no alternative courses of action; because of their long-term values or norms, their behavior changed slowly or not at all. As a result, even without the drought, a crash would have soon occurred, moving the system into chronic overgrazing and range degradation and continuing the southward encroachment of the desert that was widely believed to be occurring (70ff).

For Picardi, as for his SD teachers and colleagues, the only outside interventions or changes capable of averting the crash and downward cycle were those based on understanding the dynamics of the system as a whole. A general characteristic of SD is that agents within a modeled system who respond rationally to their local circumstances generate, through the feedback structure of the larger system, outcomes contrary to their intentions or best interests, such as vicious cycles or booms and busts. Such outcomes can be overcome only when the agents are either coordinated by a superintendent manager or transformed by a universal change of rationality. In the case of Forrester's (1961) models of firms, it was claimed a manager could solve the problem of cycles by adjusting the relations inside the firm. In *The Limits to Growth* (Meadows et al. 1972), there is no world manager, so everyone must adapt to no-growth economics. Similarly, the decentralized pastoralists are not amenable to management, so, according to Picardi's analysis, they must all adopt a range conservation ethic.

In contrast, a modeler might recognize progressive changes initiated by agents within difficult, complex situations.

> Harmsworth (1984) describes four apparently successful projects funded by a nongovernmental organization that assisted pastoralists in Niger and Mali to recover after the 1968–1974 drought and to organize for further economic change. For example, one group of nomadic herders formed an association and eventually a cooperative. In each case, the project was locally initiated or had a high degree of local participation, yielding a diversity of approaches and outcomes that reflected the local circumstances.

Modeling situations with such local particularity combines the practical difficulties of all of the previous contrasts. In short, the modeler would have to interest sponsors in detailed, long-term, locally centered research and policy. (From 1975 onward, anthropologists consulting for USAID on pastoralism emphasized exactly this; see Horowitz 1976; Swift 1979.) USAID in 1974, however, wanted a means to evaluate programs for the overall development of the region—programs that the United States might sponsor. Given that SD models consistently dictated system-scale transformations and reinforced the irrelevance of local participation in responses, Picardi had no need to renegotiate the terms of reference with the project's sponsors.

Multiple Points of Engagement

Picardi's tool was not the only conceivable one for creating a model representation of nomadic pastoralists. His socio-environmental modeling could have been modified, but, as my analysis of alternatives to the eight different aspects of SD has revealed, not without many practical consequences. Overcoming the limitations of SD with respect to, for example, restructuring (see contrast 1) would not have been a mere technical task. Picardi would have had to interest sponsors in different approaches, such as a detailed, comparative study, sustained engagement of a researcher in Africa, models applicable only in specific localities, and ongoing assessment of changes. Conversely, many interdependent resources helped Picardi to represent pastoralism as an enduring, integrated, well-bounded system. The mechanist and behaviorist language of the strong SD view of systems privileged the outside, superintending agency. This complemented the interventionist position Western nations and international bodies assumed at that time when designing policy for the development of former African colonies. USAID dictated the short study time, which limited the research and engagement that might have revealed possible restructurings of pastoralist arrangements. Picardi did not see the need to model restructuring; this choice facilitated his use of SD to represent pastoralism in clearly characterizable long-term projections. These clear projections, in turn, fulfilled USAID's terms of reference, at least, with respect to the pastoralist sector of the region.

And so on. No one resource or domain of action in this heterogeneous web stood alone: language, tools, work organization, and social relations beyond the work site reinforced one another—that is, rendered one another harder to modify. Together they constituted a stable structure: agents in the intersecting domains of action—from the pastoralists, whose voices could scarcely be heard across the Atlantic, to concerned members of Congress—were kept from disputing Picardi's scientific representations or from withdrawing the support Picardi needed to continue his work (fig. 4.6). Such cross-reinforcement of diverse resources—not some overarching technocratic rationality on the part of the researchers—forms the basis of technocratic governance.

The picture of science and scientific agents that has emerged in this chapter is more general than that of chapter 3. It can be used to interpret the social influences on scientific models when the scientists do not share H. T. Odum's particularities and visionary style. The bases of the framework I call *heterogeneous constructionism* are that

1. many heterogeneous components are linked together in webs, which implies that
2. the outcome has multiple contributing causes, and thus
3. there are *multiple points of engagement* at which the course of development could be modified.

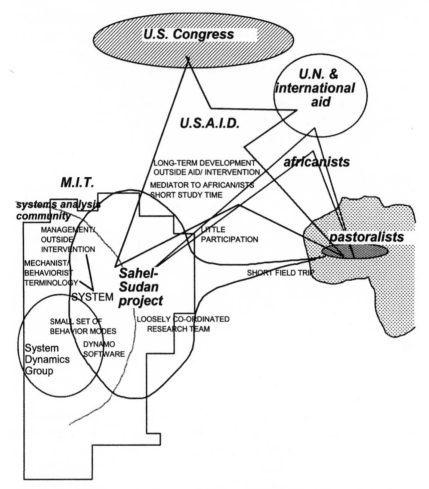

Figure 4.6 An impressionistic schema depicting diverse agents and selected resources involved in the construction of Picardi's system dynamics model. The size of the elements signifies their relative importance.

In short,

    4. causality and agency are distributed, not localized.

Teasing out this framework leads me also to note that

    5. construction is a process; that is, the components are linked over time,

    6. building on what has already been constructed, so that

7. it is not the components, but the components in linkage, that constitute the causes.

Points 2 and 5–7 together ensure that

8. it is difficult to partition relative importance or responsibility for an outcome among the different types of causes (e.g., mostly "scientific" but partly "social").

Generally,

9. there are alternative routes to the same end, and
10. construction is "polypotent" (Sclove 1995); that is, things involved in one construction process are implicated in many others. Thus
11. engagements within a construction process, even very focused ones, will have side effects.

Finally, points 5, 10, and 11 mean that

12. construction never stops; completed outcomes are less end points than snapshots taken of ongoing processes.

Within this framework, scientists in action should be thought of as imaginative agents, working knowledgeably and capably within intersecting domains of action, cross-linking heterogeneous resources over time in order to represent-engage; that is, to build, and to build on, heterogeneous webs. The outcomes of their scientific work—theories, readings from instruments, collaborations, and so on—are accepted because they are difficult to modify in practice. If we interpret science in terms of its heterogeneous construction, we need to tease out these webs of resources and expose their causal significance. Counterfactual analysis, as I have shown, is one means of exposing what makes a difference.

For Picardi and the Kerang Farm modeler, just as for Odum, the personal, scientific, and social facilitations reinforced one another, rendering it difficult to change the construction. Nevertheless, this chapter, unlike my all-encompassing interpretation of Odum, provides more fine-grained and temporally limited accounts, which reveal multiple potential sites of engagement—specific points at which concrete alternative resources could have been mobilized. If such interpretations are to bear productively on subsequent research, however, the challenge is to get scientists interested in examining their diverse resources. They need to be weaned away from assuming that the most important site for engagement is around convictions about the correctness of a certain model or theory, or its superiority over others. In a heterogeneous construction, such convictions are only one resource—they are not decisive on their own. Success in getting scientists interested in examining their diverse resources would make it possible to assess the extent to which their becoming aware of multiple potential sites of engagement helps them to alter their personal, scientific, and social facilitations, and thereby to modify the directions in which their science moves. The challenge of stimulating such heterogeneous *recon*struction motivates the explorations described in part III.

▲

My investigations of how the models in these two socio-environmental modeling projects came to be established as knowledge centered on assessment of what would be entailed in practice to modify that knowledge. By identifying alternatives to specific aspects of these modeling projects and teasing out their practical implications, I was able to trace diverse interconnections between the various so-called technical tasks of scientists and the social considerations that influence how scientists perform these tasks. The terms and themes I formulated emphasized that scientists harness many diverse resources in establishing knowledge. This process of heterogeneous construction is always, in practice, bound up with construction of lives, careers, institutions, language, ideologies, societies—that is, with a range of actions and engagements. Scientists are simultaneously representing and engaging. In this sense, the work of modelers embedded in a social context became a variant of intersecting processes (see the narrative at the end of chapter 1 and chapter 5, section C)—a variant whose interpretation requires special attention to the agency of the modelers.

At this point, I saw these scientists' agency as something distributed beyond their persons, depending on webs of resources, such as the available computer software, published data, the length of study time set by the sponsors, and so on. This view, which extends the themes of heterogeneity and embeddedness (see the prologue and chapter 1), contrasts with the idea of agency as something concentrated inside scientists' minds in the form of motivations, beliefs, perspectives, biases, or ideology. Concentrated agency steers attention toward verbal and textual discourse, but I was choosing to emphasize the diverse material aspects of practice relevant to constructing knowledge (see notes 11, 12, 14, 15, and 30).

The shift I had made away from overall correlations between scientific content and social-personal context to the picture of heterogeneous construction had been accompanied by three other shifts of emphasis. The first was that the ecological theorizing discussed in part I had given way to a focus on research on socio-ecological complexity—first Odum's "systems of man and nature," then projects of socio-environmental assessment.[18] I had not left my interest in ecological theory behind; indeed, this new focus had provided material in which I could explore not only the *sociality of science, but also the problematic boundaries of ecological or environmental complexity. My analysis of alternatives to system dynamics modeling of nomadic pastoralists had acquainted me with the field of political ecology, in which cases of environmental degradation were explained by linking local changes in agroecologies and in labor supply and the organization of production with wider political-economic conditions (Peet and Watts 1996b). In short, this was an area giving substance to Wolf's image of intersecting processes that involve diverse components and span a range of spatial and temporal scales (Wolf 1982, 387; see the narrative at the end of chapter 1).*

I noticed some affinity between this first shift and steps being taken by ecologists developing an approach called Adaptive Environmental Management (AEM) (Holling 1978). AEM promoted use of multiple models and their ongoing revision in recognition that any ecological situation is a moving target—not the least because management practices produce continuing changes. In my terms, AEM was addressing the ongoing restructuring and embeddedness of ecological situations. (Subsequently, AEM has evolved into a field that advances models of the social or institutional embeddedness of research and policy.[19]) At that time, however, I was less enthusiastic about AEM's orientation toward environmental management. My critical perspective on the technocratic orientation of the socio-environmental research of Odum, Picardi, and the Kerang Farm study (chapters 3 and 4) motivated a second shift in emphasis; namely, to look for examples of representing and engaging in ecological situations that were less technocratic—cases in which researchers bridged the divide between outside analysts and the subjects whose social and ecological situation was being analyzed. In this vein, I was inspired by cases of participatory action research (PAR), in which the researchers shaped their inquiries through ongoing work with and empowerment of the people most affected by some social issue (Adams 1975; see the epilogue, sections A and B).

The final shift of emphasis was that assessment of what would be entailed in practice to modify the knowledge produced by the modelers primed me to reflect on the social considerations that shaped my own research as an interpreter of science. My efforts at self-conscious or reflexive engagement with ecological and social complexity, which is the subject of part III, were informed by the PAR ideal ...

▲

Part III

Engaging Reflexively within Ecological, Scientific, and Social Complexity

One motivation for the efforts in interpreting the sociality of science presented in part II was the possibility that awareness and discussion among ecologists of such interpretations might influence their subsequent work in productive ways. It was still an open question how best to feed interpretation back into science. On one hand, interpreting two socio-environmental projects (chapter 4) had led me to identify alternatives to specific aspects of the modeling methods employed in each case. These alternatives pointed to the possibility of representing complexity without assuming the existence of well-bounded systems, an assumption that my earlier modeling and historical work had also called into question (chapters 1 and 3). I knew that scientists—including the scientist in me—would like to see what modeling built around the alternatives would look like. On the other hand, interpreting those modeling projects had also heightened my awareness of the diverse practical considerations and interactions among diverse social agents involved in establishing what counted as knowledge. I decided, therefore, not simply to focus on ways in which ecological and socio-environmental complexity could be modeled so as to capture ongoing restructuring, heterogeneity, and embeddedness. I also needed to explore the potential of heterogeneous constructionist interpretations to expose many specific sites of scientific practice at which different researchers—interpreters of science as well as scientists—could engage with a view to modifying the science (chapter 4). I would need to get more interpreters of science interested in analyzing scientists' diverse resources and making sense of their distributed agency. I would also need to get scientists to pay attention to such interpretations—or, even better, to become interpreters of the construction of their own work; that is, to become practically reflexive. And I would need to follow through on the implications of such reflexivity, which included modeling what I wanted for scientists in the ways I personally interpreted and engaged with science …

5

Reflecting on Researchers' Diverse Resources

I had developed my interpretation of the computer modeling of nomadic pastoralists in the context of making a contribution to a set of sociological papers on the tools used by scientists. In the course of this work I had already begun to explore ways to engage others in analyzing scientists' diverse resources (section A). During the same period, I organized some workshops in which researchers reflected explicitly on their own sociality and how it affected their work, and were encouraged to identify for themselves potential sites of engagement and change (section B). Soon after, I took up a position teaching about biology and environmental science in their social context. This position provided more opportunities to attempt to distribute the work of interpretation and engagement to others (section C) ...

◆ ◆ ◆

A. Further Intersections that Affect Researchers and Interpreters Extending Their Webs

What do interpreters of science do to produce the "right" account for influencing their audience?

In 1987, I had set myself the task of critically reconstructing the work Picardi did in constructing the models of pastoralists and their resources. Originally there were two intended audiences: socio-environmental modelers and interpreters of science.

An invitation to contribute to a symposium in 1989 on the "Right Tools for the Job" turned me toward the latter group (Clarke and Fujimura 1992a). The sociologists of science who initiated the "Right Tools" project were interested in shifting attention from scientific ideas and theories to various aspects of the everyday practice of science, including materials, tools, language, work organization, and wider social relations. They proposed that sometimes the choice of a particular tool or organism allows many aspects of practice to reinforce one another—to be "aligned" (Fujimura 1987; Clarke and Fujimura 1992b, 8)—and accelerates the pace of the science. In such situations, the tool or organism was the right one for the job.

To address the themes of these sociologists, I identified Picardi's tool—system dynamics modeling—and adopted it as the focal point for my analysis. But in what sense could I argue that system dynamics was the right tool? From my initial research, I already knew that the web Picardi wove around his modeling had not been extended to other projects—his tool was not right enough for that to happen. It seemed right, however, to the extent that the web stabilized the intersecting domains of action or "social worlds" (Clarke 1990, 1991; Fujimura 1988). The claim I decided to make in the context of the Right Tools project was that for Picardi at MIT in 1974, many practical considerations reinforced one another so that modeling differently would have been very difficult.

Cross-reinforcement provided a way to connect with my audience of interpreters of science around the theme of the rightness of tools, but I saw complications arising. The rightness of system dynamics—the difficulty of Picardi modeling differently—could not stand or fall on whether he was personally motivated to alter his modeling, or on whether he ever attempted to do so but encountered resistance. After all, social agents are routinely facilitated or constrained by factors or relations that they consider unproblematic or take for granted. To expose such factors, however, raised difficulties in both analysis and presentation.

The first set of difficulties stemmed from the counterfactual approach I used to identify resources. This approach drew me away from established practices and conventions in history and sociology of science. These fields favor representing how agents in their own time and setting have seen their situation. The narratives in these fields, despite their implicit causal claims and counterfactuals, tend to confine the author to what happened and omit discussion of what did not happen but might have. Recognizing the tension between abiding by the agents' own accounts and my interest in exposing unspoken factors, I decided to begin in the familiar narrative mode, but then to interrupt it. I would observe conventions in order to capture the attention of historians and sociologists and orient them to my project, but then I would attempt to get them interested in disturbing those conventions and in entertaining a more causally explicit counterfactual analysis.

Shaping my work to fit the Right Tools theme led to a second set of problems. In the course of my research, I tried to account for why the web Picardi had woven had not been extended after the completion of the MIT project. The additional detail I uncovered indicated how Picardi's web was destabilized—and how my reconstruction could suffer an analogous fate. Each term—jobs, tools, and rightness—became problematic. (Although I did not know it at the time, other members of the Right Tools project were also reaching this conclusion, which eventually led to a loosening of the Right Tools theme; Clarke and Fujimura 1992a.) I decided to turn my deviation from the original theme of rightness into an opportunity. I followed the reconstruction of Picardi's modeling (given in section B of the previous chapter) with a postscript, in which I aimed to provoke discussion among the Right Tools participants and their audience of certain conceptual and methodological issues. Those issues were obvious in the analysis of heterogeneous webs, but relevant, I wanted to argue, to all concerned with the complexities of everyday practice of science. My hope was that, if interpreters of science concerned themselves with conceptual and methodological issues involved in addressing complexity, this would draw their work closer to mine. The overlap would enhance the plausibility of web analysis, even for members of my audience who were not moved to adopt my counterfactual approach. And, in the long term, the more common web analyses became, the more plausible scientists might find the ideal of scientific practice informed by web analysis. The effect I hoped for, therefore, would depend less on the immediate acceptance of my interpretations as representations of socio-environmental modeling than on the webs I could build supporting them.

I present that postscript here, first narrating what followed the completion of the MIT project, then reflecting on the conceptual and methodological challenges of reconstructing heterogeneous webs. (Where appropriate, the original postscript is expanded to include references to the reconstruction of the Kerang modeling study presented in chapter 4, section A.) In an afterword, I reflect on what I did not raise in the original postscript.

Postscript to the Reconstruction of the MIT Modeling Study, Part 1: A Web's Destabilization

Given that the components of Picardi's heterogeneous web were, in my interpretation, mutually reinforcing, it is reasonable to ask how long that

web remained stable and whether it was extended to other projects. In the
fall of 1974, USAID extended the MIT contract for three months so the
reports could be completed, but no additional funds were allocated.
(Picardi completed his report within the original timetable and used the
extension to develop models that included some economic and social motivations for pastoralist decisions.) MIT's requests to be funded for follow-up
research were turned down. Seifert, the project leader, was commissioned
by the Saudi Arabian government to direct an engineering study of their
water systems, and Picardi joined this project. Meanwhile, USAID selected
two components of the MIT study—a summary report and the framework
for planning agricultural development—for distribution to international
organizations and African governments. The remaining ten studies, including Picardi's, were packaged as "annexes," available only upon request.
The report received little reaction, and no requests came for translation to
allow wider dissemination in predominantly francophone West Africa.

In June 1975, USAID sponsored a colloquium in West Africa on the
effects of drought on herders and farmers (Horowitz 1976), in which no
members of the MIT team participated. The first of the meetings of international donors and African representatives proposed by the U.N., for which
the study was initially produced, did not take place until the inaugural
meeting of the Club des Amis du Sahel in March 1976. In a background
document for that meeting, the MIT project was one of several long-term
development studies reviewed, but was cited specifically only once. Ironically, the citation was to the study's rejection of advanced technologies as
the solution to the region's problems (Giri 1976).

In April 1976, USAID reported to the U.S. Congress on its "Proposal for
a long-term comprehensive development program for the Sahel" (USAID
1976). Despite the title, this report was based not on the MIT study, but on
the efforts of the Sahel Planning Task Force, a new group of consultants
and USAID employees convened in 1975. Components of the MIT report
were cited in the supporting documents, but the proposals bore no mark of
MIT's systems analyses. In particular, the consultants' report on pastoralism
conveyed a positive view of herdsmen as "flexible," "innovative," and able
to use an "otherwise economically sterile zone" (USAID 1976, part II,
47–50). (The author of this section appears to have been the task force's
consultant on anthropology, Michael Horowitz, a researcher with extensive
field experience in the region and a critic of development policy imposed
from afar; see Brokensha, Horowitz, and Scudder 1977.)

USAID's creation of the Sahel Planning Task Force and its failure to promote MIT's efforts might suggest to us that the completed MIT research

project was not a resource for USAID, either in 1975 or subsequently. What brought about this state of affairs? How does it reflect on the strength of the cross-reinforcements within Picardi's heterogeneous web? When I explored the background of the USAID-MIT parting of ways, I was led to a more complex picture of USAID-MIT relations during the course of the Sahel-Sudan project.

In the fall of 1973, the National Academy of Sciences (NAS) had convened a panel on "Arid lands of sub-Saharan Africa" to advise USAID on natural resource management and U.S. assistance (NAS 1975a, 1975b). The panel never clearly defined a mission distinct from MIT's, but USAID asked it to review the progress and reports of the MIT study. These reviews were increasingly critical, and USAID encouraged panel members to develop outlines of alternative frameworks for analyzing the region's problems (NAS 1975a, 1975b). USAID internal documents reveal that, as early as the month after the MIT contract was signed, some doubts were expressed about MIT's ability to produce the internationally recognized analysis USAID wanted, and by March 1974, USAID requested a major restructuring of MIT's work plan (Lyman 1974; Hobson 1974; Glantz, Robinson, and Krenz 1980). MIT, however, was able to resist any significant changes in personnel or project organization. USAID did not withdraw funding, but began to distance itself from the project, anticipating that it would not be the resource they desired for guiding U.S. and international assistance to West Africa. In October 1974, when the NAS advisory panel's efforts for USAID culminated in a conference on "International Development Problems in the Sahel," no one from the MIT team was present.

The NAS and USAID documents indicate a steadily deteriorating relationship between USAID and MIT. They also complicate the earlier account of the different domains of social action that intersected in the project. A new agent, the NAS panel, takes a position as one of USAID's resources. In addition, USAID's own internal tensions begin to be revealed in the documents. USAID was not a single undifferentiated phenomenon; namely, an agency expressing clear needs that MIT simply did not meet. Instead, I found that USAID was "torn in both directions." One road to development was to intensify productivity using technical and infrastructural improvements; the other was to provide "widespread small-scale assistance to farmers and herders based on their perception of their needs" (NAS 1975b, 4; see also USAID 1975, 1976). The MIT study had not provided the science- and capital-intensive plan USAID administrator Adams had originally envisaged and, if anything, had leaned toward the small-scale alternative.

During the course of the project, however, Picardi's construction of a heterogeneous web around his modeling was not destabilized by the instability of the relationship between MIT and its sponsors. In fact, although a USAID review of the Sahel project's progress after five months was generally critical, Picardi's livestock modeling was described as "one of the most original and promising approaches in the report" (Lyman 1974). Furthermore, after the final report on the Sahel project was released, I did not find a rejection of Picardi's particular conclusion that the pastoralists were caught in a vicious cycle of overstocking and range degradation. Nor was the SD methodology for addressing socio-environmental problems subjected to critique.

Some members of my audience might consider that Picardi's job—building models that were acceptable representations of reality—ended when he was awarded his doctorate. If so, then the post-study non-ramifications of the MIT study and the more complicated MIT-USAID-NAS relations become irrelevant. However, Picardi was not simply a doctoral candidate; after six years of experience at MIT in systems analysis of environmental issues, he fully expected to continue applying SD to socio-environmental problems. Nevertheless, Picardi recognized even before the Sahel project's completion that support was lacking for extending his work in this area:

> Further research on this tragedy-of-the-commons problem will be useful only if it enlists the participation of policy-makers responsible for long-term decisions in the Sahel and of experts with intimate knowledge of pastoral and ecological dynamics. . . . No further analytical frameworks will be useful unless the public officials responsible for decisions feel that these frameworks will help them make decisions on their long-term problems. Therefore, the direction of possible and *useful* future research simply cannot be stated at this time. (Picardi 1974c, 219; his emphasis)

Picardi had noticed that USAID had not supplied needed resources to, or had withdrawn them from, the MIT project (Picardi 1974a). His models were not disputed by USAID as representations of pastoralism. Yet, within the enlarged picture of USAID-MIT relations, any chance that Picardi could extend his heterogeneous web was precluded by his belonging to a research organization that was unable to stabilize, let alone extend, the intersection of domains of action that had brought the Sahel project to MIT.

Postscript, Part 2: Challenges in Web Analysis

By delving into relations between USAID and MIT, I have expanded the web of agents and resources implicated in Picardi's modeling. Picardi's modeling work, moreover, becomes just one possible entry point—and not necessarily the best—for reconstruction of the larger heterogeneous web of scientific activity. My exposition, however, has not proceeded directly to this assessment. Instead, it has followed the steps of my reconstruction, first presenting a strongly cross-reinforcing web, and then destabilizing that reconstruction. This approach allows me to raise a number of questions, starting with the choice of entry point and of boundaries for the web. In outlining those questions, I will indicate the different methodological choices that could have been made, at least in principle. I believe these concerns are relevant to any interpretation that attempts to make sense of the diversity of resources drawn on in the everyday practice of science.

1. **Entry point.** Why focus on how support was constructed for Picardi's models—or, similarly, the Kerang Farm Model—as representations of reality? Why not focus on how the modelers managed to conduct their everyday work in their institutions? Indeed, why choose to probe the webs from the vantage point of the modelers? Why not, instead, enter the webs for the MIT project around the concerns of USAID, which was trying to use the MIT study to elevate the role of the United States in West Africa, traditionally the province of the French?

2. **Boundaries.** Who and what have I included within the intersecting domains of social action? Who and what are excluded or located outside the web? Without shifting the interpretive focus away from the modelers, my analyses could be expanded so as to examine the impinging construction of webs by the other agents most obviously implicated in the studies: in the MIT study, Seifert and USAID; in the Kerang study, the principal investigator, the agricultural economist, the Ministry, the agricultural extension officers, and the farmers. We could probe even further, given that these primary agents built on the webs of other agents. The pragmatist would say that we cannot do everything, so one has to be selective, but selectiveness also implies that excluded webs do not make a significant difference. To show this, we need to explore transgressions across boundaries and web construction that reticulates or reverberates to other webs. In this spirit, even if we set out to account for how scientists supported their theories and representations, we could end up interpreting their scientific activity more generally.

Once we begin to analyze the construction of heterogeneous webs that connect in a reticulating or reverberating manner to other webs, more questions arise:

3. Categories. In what terms should we economically describe the agents, their resources, and their interactions? For example, if the boundaries are drawn so that the web includes research sponsors and the domains of political action in which they work, categories for analyzing their organization and activity are needed. Moreover, once we admit the possibility of resources that may not be articulated by the agents, there is no good reason to limit ourselves to the categories the agents use. Picardi, for example, strongly believed that if the feedback dynamics within systems were not understood, people's apparently rational actions might generate undesirable long-term outcomes. I have chosen, however, not to explain his models in terms of those beliefs, but to go behind them and analyze the heterogeneous web Picardi was enabled to construct.

4. Systemization. What order can be abstracted from the complexity of interactions and interconnections? When and how can the webs be partitioned into levels, distinguishable according to the scale and extent of the actions involved? The cases of chapter 4 show that, even when I focused close to the tool-using sites, the resulting interpretations included both immediate and less direct resources, and both the general and the particular circumstances.

5. Particularity and contingency versus generality and determinism. Do we emphasize the particularity and contingency of agents' actions, or do we generalize about agents despite differences in their situations? If we seek generality and discount particularity, will the outcomes appear simply determined? Can we acknowledge particularity and contingency, yet remain open to identifying regularities and structure?

6. Integration versus change. Do we look for interactions that maintain the web's integration or structure and its adaptation to the "external" context? Or do we look for sources of change and restructuring and places where the web is least resistant to modification? Can we talk about structure without discounting sources of destabilization and change? Similarly, can we talk about closure, agreement, or understanding without discounting their provisionality? When I paid attention to the heterogeneity and change in the intersecting domains of action, it became more difficult—as indicated by the narrative of the aftermath of the MIT study—to characterize a scientist's work as one job, or to claim that the right tool was applied.

◆ ◆ ◆

There are no easy answers to these questions, which *in principle* face all interpreters making sense of situations in which scientists and other agents use a diverse and distributed set of resources. *In practice,* of course, choices are made. For example, I made choices that can be summarized in terms of a self-conscious engagement with the science. Specifically, I wanted my interpretation of Picardi's modeling to contribute, albeit indirectly, to participatory rather than technocratic approaches to socio-environmental studies (see the narrative at the end of chapter 4; this chapter, section B; chapter 6, section C; and the epilogue). A different interpreter could have moved in the opposite direction. Recall that Cockrum, one of Picardi's advisors, wanted Picardi to concentrate on the livestock-pasture relationship and omit social and economic relationships altogether (see contrast 2 in chapter 4, section B). Given my interest in participatory approaches, however, I did not frame my counterfactual analysis around the purely agronomic modeling that Cockrum advocated. In a similar spirit, I chose to highlight Picardi's web of resources rather than his internal motivations or beliefs. This choice corresponds to my interest in exposing multiple potential sites of engagement so that I might help subsequent scientists in comparable situations to mobilize alternative resources.

An engagement with the science is implied—even if I were not promoting participatory approaches—because my reconstruction of Picardi's modeling goes against his self-conceptualization of SD. Picardi believed, for example, that science requires defining one's questions, dividing the world accordingly into a system and its context, and capturing that context through the system's boundary conditions. As he reaffirmed in an interview, "that's the core of the scientific method" (Picardi 1989b). Clearly I have been contesting that conception of science.

Historians and sociologists are also engaging with the science when they make their accounts faithful to the scientists' and other agents' own voices and eschew counterfactuals—what might have happened—in favor of what did.[20] Such accounts privilege relations that those agents considered unproblematic or took for granted. Just how far, and in what directions, we dispute the taken-for-granted is, I contend, central to reconstructing the web of resources built into scientific work. For example, in her multifaceted account of the oncogene bandwagon, Fujimura, like the agents she observes, does not remark on the prevailing emphasis on curing cancer over preventing it (Fujimura 1988). This socially contingent commitment is surely a resource for those scientists; taking it for granted implies an assessment that the "costs" of disputing it are too high. Socio-environmental modeling appears to be a case in which heterogeneous webs of research are more readily destabilized. But, as I have tried to demonstrate, destabilization is always an option. Like system dynamics in Picardi's hands and then in mine, any tool that appears right—in fact, any job, theory, or institution in science—is open to reconstruction.

Afterword: Practical Reflexivity

In concluding this postscript, I played off the work of one of the Right Tools organizers, Joan Fujimura, implying that conceptual and methodological issues should be of concern to the other members of the Right Tools project. I was trying to open up questions, but not to go further and spell out how I thought those questions connected with interpretation centered on heterogeneous construction. Interpreters of science would have to address the complexities of sociality for themselves, I thought, if my framework of heterogeneous constructionism were to become something they would work with.

In principle, however, I could have elaborated on my conclusion in the following terms: Interpreters of science are, like scientists, always representing-engaging. In dealing with heterogeneous webs—even if only to discount their complexity—interpreters of science also build their own webs. They select and juxtapose components in narratives, fashion boundaries and categories, employ conventions of representation, adopt and adapt current theoretical themes, and so on. They want to convince intended audiences, secure ongoing support from colleagues, collaborators, and institutions, and enlist others to act on their interpretations—or, more broadly, to stimulate others to build webs that reinforce their interpretive work.

To be consistent, interpreters of science should then allow their interpretations to be interpreted—perhaps doing so themselves. This would mean that web analysts would reflect on the full range of practical conditions that enable them to build and gain support for their own representations. In short, they would be *practically reflexive*.[21] (As an analogy, recall that the modeler in the Kerang study considered living and working in the Kerang region as an agricultural consultant, adjusting his models to reflect sociologically realistic processes of how farmers change. This possible move raised a range of practical issues, from the purchase and maintenance of a car to keeping abreast of discussions about the wider state of the rural economy.) If interpreters were practically reflexive, they would model for scientists the process of reflecting on one's diverse resources with a view to modifying the directions in which one's research moves.

This call for practical reflexivity was clearly foreshadowed in the personal, reflective tone adopted at the start of the chapter. The postscript, however, did not take practical reflexivity far. Instead of unpacking the diverse practical considerations that led to my own choices, I invoked generalities about engaging with the science and promoting participatory approaches; there was none of the detail I had provided on Picardi's work. Similarly, I played down the practical considerations members of my audience might face in addressing the conceptual and methodological issues raised.

However, it was not merely inconsistency that led me to show less practical reflexivity than was conceivably possible. I realized the difficulty—as a practical and an expository matter—of examining the full range of practical conditions relevant to gaining support for interpretations of science. I also observed that most historians, sociologists, and other interpreters of science show little interest in practical reflexivity and in examining the engagements facilitated by their interpretations. Conventionally, even when the practical dimensions of science have been emphasized, the practical dimensions of interpreting that science have not. I decided for the most part to observe this convention and reserve my convention-disturbing chips for the move from narrative to explicit causal analysis. Finally, in turning attention to the interpreters of science, I was less interested in making interpreters more reflexive than in encouraging them to *take up* web analysis of scientists' work. Recalling the conclusion of chapter 4, I did not think that presenting the particular complexity of my own web of resources would help other interpreters to "alter their personal, scientific and social facilitations, and so modify the directions in which their [interpretation] moves." The best I thought I could do was to stimulate interpreters to take up web analysis; I would leave it to them to mobilize the resources they needed to do so. This was my intention in presenting the conceptual and methodological issues as *logical* consequences of acknowledging—as most of the Right Tools participants did—the existence of many, diverse resources or elements in the practice of science.

Soft-pedaling the practical dimensions of interpreting science amounted to a particular choice of boundaries and categories on my part. Indeed, there is nothing natural about interpreters leaving practical reflexivity in the background; it cannot be a matter of simply representing reality faithfully. Such a justification ceases to apply once interpreters acknowledge that, like scientists, they mobilize multiple, diverse resources in establishing what counts as knowledge and in pursuing their work more generally. Each of the methodological issues identified in the postscript exposes the hidden complexity of simple formulations (recalling chapter 1, section B) that omit the situation of the researcher.

Clearly, there was a tension left unresolved in the postscript. I wanted my audience to recognize the additional layers of complexity that follow from interpreting science as heterogeneous construction. Yet, as a matter of practical reflexivity, I chose to abide by simpler expository conventions with which the audience would be comfortable. This meant that I did not analyze the full range of practical conditions relevant to gaining support for such an approach to interpreting, or representing-engaging with, science. At the time of the Right Tools project, however, I was also pursuing a path that might resolve some of this tension.

B. Workshops in which Ecologists Map Their Webs of Knowledge-Making

Can scientists become interpreters of science and bring their interpretations to bear on their science?

According to the theme of distributed agency (chapter 4, section A), scientists mobilize a diversity of resources and, in so doing, engage with a range of social agents. According to the postscript to my reconstruction of Picardi's modeling project (chapter 5, section A), when interpreters of science delimit the relevant resources and agents, they also mobilize resources and engage with diverse social agents. Interpreters who recognize this might then reflect explicitly on the practical conditions that enable them to build and gain support for their interpretations. Applying the same interpretive framework to their own research should enhance the plausibility of their reconstructions of the work of scientists.

There might be a more direct way for heterogeneous constructionist interpretation to influence science productively. Instead of relying on some second party to do the reconstruction, could scientists—or indeed, any researchers—interpret their own heterogeneous webs? Could researchers reflect explicitly on how their own social embeddedness or situatedness affects their ability to study the situations that interest them? Could they attempt to identify multiple potential sites of engagement and change for themselves? If so, this strategy would cut through some of the complexities arising from interpreters trying to model practical reflexivity.

Mapping, Mapmakers, and Maps

To explore this possibility with a number of ecologists and natural resource researchers, I convened two "mapping workshops": the first in Helsinki, co-led with ecologist-philosopher Yrjö Haila; the second in Berkeley.

These workshops were designed to proceed as follows: Each researcher would focus on a key issue—a question, dispute, or action in which the researcher was strongly motivated to know more or act more effectively. All researchers would identify "connections"—things that motivated, facilitated, or constrained their inquiry and action. These might include theoretical themes, empirical regularities, methodological tactics, organisms, events, localities, agents, institutional facilities, disputes, debates, and so on. Researchers would then draw their "maps"—pictorial depictions employing conventions of size, spatial arrangement, and perhaps color that would allow many connections to be viewed simultaneously. The map metaphor was meant to connote not a scaled-down representation of reality, but a device that shows the way—a guide for further inquiry or action (Taylor and Haila 1989; Taylor 1990).

Over a series of sessions, the workshop participants would present these maps and be questioned by other participants. As a result, they might clarify and filter the connections and reorganize their maps so as to indicate which connections were actually significant resources. The ideal was that researchers would self-consciously modify their social situations and their research together, perhaps in collaborations formed among the workshop participants. Of course, given that the mapping was an experiment, it was not surprising that the ideal was not realized in these initial two workshops.

Three maps from the workshops illustrate the map making that resulted. Figure 5.1, by a Finnish ecologist I will call "E," was the most orderly of the maps, having been streamlined and redrawn on a computer. As such, it does not do justice to the real-time experience of its production during an actual workshop. Indeed, when viewed on their own, all the maps appear schematic; valuable history, emphasis, and substance were added when the mapmakers presented their maps to other workshop participants.

The central issue of E's map is very broad; namely, to understand the ecology of carabid beetles living in the leaf litter under trees in urban environments. On the map below this issue are many theoretical and methodological sub-problems, which reflect the conventional emphasis in science on refining one's issue into specialized questions amenable to investigation. Above the central issue are various background considerations, larger and less specific issues, situations, and assumptions that either motivated work on the central issue or were related to securing support for the research. E's research alone would not persuade the urban public to recognize that "nature is everywhere—including the cities," but by combining the upward and downward connections shown in the map, he reminded

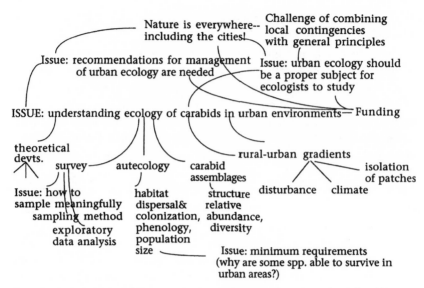

Figure 5.1 Redrawn outline of E's map on how to conduct research on the ecology of carabid beetles in the city of Helsinki. (From Taylor and Haila 1989.)

himself that work on the background issues, as well as refining a working hypothesis, would be necessary to be able to keep doing his research.

In narrating his map, E mentioned some additional history. Many of the ecologists with whom he collaborated had been studying a forest area, but the group lost its funding when the Forestry Department asserted that forest ecology was its own domain. It did not matter that animals are barely mentioned in the ecology of forestry scientists. The ecologists self-consciously, but of necessity, turned their attention to the interconnected patches of forest that extend almost to the center of Helsinki, and they explored novel sources of funding and publicity, including a TV documentary. The upward connections were thus a recurrent, if not persistent, influence on E as he defined his specific research questions.

Historical background depicted in a narrative format is more evident in a large map by "R," a Mexican researcher who had come to specialize in the economic and agronomic dynamics that lead to impoverishment of peasants, their migration into forest areas, and subsequent clearing of those forests. Figure 5.2 is only one section of that map. Although radically different from E's redrawn map, R's map also highlighted simultaneous issues of building the disciplinary and collaborative context in which to pursue his many concerns. As a biologist, he wanted to stem rainforest destruction; as a political activist, he wanted to reduce rural poverty; and as a resource

Figure 5.2 Extract from R's map concerning research on the peasant economics and politics involved in tropical forest destruction in Mexico. (From Taylor 1990.)

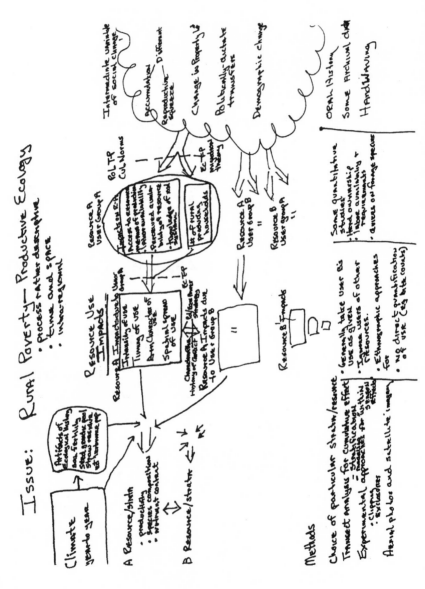

Figure 5.3 M's map of his research into ecological degradation and impoverishment among nomadic pastoralists in West Africa. (From Taylor 1990.)

economics graduate student in the United States, he needed to frame technical questions that could be answered.

In figure 5.3, "M," an American researcher studying land degradation
and impoverishment among nomadic pastoralists in West Africa, depicted a
more conventional conception of research. Questions form the bulk of the
map and are separated from methods—the strip along the bottom. M omitted the movements, arrangements, alliances, and negotiations he built in
order to monitor milk production, elicit from the herders rules governing
herd movement, assess herd ownership, measure the effect of grazing on
pasture growth, complete surveys to "ground truth" satellite images, and so
on. M's map also located him in his remote field area, and omitted the audiences in the United States—sponsors and critics alike—for his current and
future research. In short, notwithstanding the guidelines I had given to
mapmakers, M included the situation he studied and left himself out.

To what extent, recalling the goals of mapping workshops, did the
workshops lead participants to "clarify and filter the connections and reorganize their maps"? It took considerable time to prepare the maps, and
the mapmakers did not devote more time to redrawing their maps in
response to interaction during the mapping sessions. M, for example, did
not redraw his map to include his own context. To what extent, then, did
researchers realize the ideal of "self-consciously modify[ing] their social situations and their research together, perhaps in collaborations formed
among the workshop participants"? Several participants, at the Helsinki
workshop in particular, claimed that the mapping workshop had expanded
the range of influences, both theoretical and methodological, that they
would bring into planning their future work. One workshop participant
commented that mapping made it impossible "simply to continue along
previous lines." Nevertheless, although the workshops provided the opportunity to link up with others around revealed affinities, no new coalitions
emerged; changes in the researchers' work were not so dramatic.

Reflecting on Extending Mapping Workshops

Although the goals of the mapping workshops were not fully met in these
initial experiments, several lessons can be drawn for the more general project of helping researchers to reflect on their situatedness and to act self-
consciously to change their subsequent scientific practice:

1. **The workshop participants were self-selected and were by no means
representative of researchers**. Almost all of them were advanced graduate

students willing to commit time to reflecting on their research and possible future directions. Having cut their teeth as researchers, they were now receptive to expanding the range of influences, both theoretical and practical, involved in planning their work.[22] The challenge for a workshop convener is to attract researchers other than students and to sustain their interaction long enough for maps to be revised and new collaborations to emerge. On a simple level, map revision could be helped by computer software suitable for drawing and redrawing maps, so that researchers would be better able to respond to the input of other participants in the workshop. At a more fundamental level, workshop conveners who hoped to achieve wider participation and sustained interaction would need more institutional resources, workshop-leading skills, and time than Haila and I had during these initial mapping workshops.

2. Mapmakers may not be successful in modifying the directions in which they subsequently move. The original assumption behind mapping was that identifying multiple potential sites of engagement would help mapmakers change, but a successful outcome does not necessarily follow. As it turned out, for example, E was not able to complete his study of urban carabid ecology. Making a map or producing some other account of how research is constructed provides no guarantee that researchers will become able to mobilize different resources to their advantage. Stanley Fish, an influential interpreter of legal texts and literature, takes this insight a step further and asserts that reflection on one's situatedness is *irrelevant* to changing it (Fish 1989).

Not surprisingly, given my view that all scientific agents "assess . . . the practical constraints and facilitations of possible actions in advance of their acting" (see theme 3 in chapter 4), I dispute Fish's assertion. It should be an empirical matter—one to be established through experiment and experience—which kinds of reflection, workshop processes, and modes of interaction and support contribute most to scientists modifying and restructuring the situations in which they undertake research (see chapter 6, sections C2 and C3, and the epilogue). Of course, would-be workshop conveners who hope to experiment and apply the experience gained would need resources, such as those to which I alluded in the previous paragraph.

3. The maps were centered on the individual mapmaker, tended to be idiosyncratic, and were not explicit about theory about the researchers' situatedness in society and its implications for their scientific practice. Again, further experiment and experience would be needed to promote more systematic map-making approaches and to assess their value. What might happen if, say, workshop leaders urged a standard format, offered models from analogous situations, or promoted various theories or propositions about micro- and macro-social change (see section C and note 27)? Would some idiosyncrasy still have to be encouraged to ensure that scientists reflect freely on and consider changes in their own particular research settings?

Deciding on the extent to which to seek regularized, theoretically explicit maps recalls all the conceptual and methodological choices identified in section A of this chapter. The *boundaries* of maps call out for negotiation—how far away from the individual researcher should the "horizon" of the map be drawn? Should something other than the researcher's issue be placed at the center? If shifts in focus are entertained, the appropriate *categories* for interpreting and engaging with science are far from obvious. The traditional focus that scientists and philosophers place on scientific claims can be stabilized only by separating research questions from research work and social support. These realms are routinely traversed by scientists, however, even as they talk as if their scientific work derived only from the situations studied, not from their situatedness. Moreover, once mapmakers acknowledge the existence of resources in their work situation or in the wider social context, should they look for regularities or structure in those resources? Should they borrow from social theory and attempt to *generalize* about the situations in which research is done? To the extent that generalizations discount or filter out the contingency and idiosyncrasy of scientists' actions, do they inject a degree of determinism not apparent in the individual situations? Finally, as answers to these questions are decided, mapmakers might ask what engagements or *social actions* they are privileging and facilitating.

These questions could be posed not only to mapmakers, but also to anyone "reflecting on their situatedness" with a view to "acting self-consciously to change their subsequent scientific practice." Yet note the tension between pragmatic considerations and the logic of posing these as open questions (recalling here the discussion of practical reflexivity in section A of this chapter). It is quite a challenge for mapmakers to choose and depict the diversity of connections around their focal issue, without the additional task of reflecting deeply on the categories, boundaries, generalizability, and so on, of their maps. Indeed, future workshop leaders may facilitate mapmaking by supplying a template of interpretive categories and themes. Even so, there is nothing natural about the depth to which mapmakers (or other researchers) reflect on their situatedness in seeking to change their practice. Any choice that mapmakers make or take for granted could be queried by their fellow workshop participants, who could ask how readily that choice could be modified. Such probing would begin to expose diverse practical considerations that support such choices.

◆ ◆ ◆

Mapping workshops offer a more direct path than interpreters of science modeling practical reflexivity (section A of this chapter) for bringing interpretation of science's sociality to bear productively on scientific practice. But they do not escape practical reflexivity's tensions and complexities, which remain not only for the mapmakers, but also for workshop conveners. Conveners might want their workshops to distribute among others the work of interpreting and engaging with research, but this goal is unlikely to be realized unless the conveners have significant institutional resources, workshop leading skills, and free time. Whether such resources can be assembled is a matter not of the workshop conveners' will alone, but of their distributed agency. In my own case, while waiting for an appropriate conjunction of circumstances for further mapping workshops, I used teaching and scholarly presentations to pursue a third approach to encouraging researchers to reflect on their diverse resources.

C. Two Terms that Help Researchers Conceptualize More Complexity

Could something simpler than mapping crystallize the need for researchers to mobilize different resources or organize them in new directions?

In mapping workshops, researchers reflect explicitly on how their own situatedness affects their ability to study the situations that interest them. The ideal is that such reflection allows them to identify potential sites of engagement and change that they had not previously entertained. Notice, however, that from the perspective of heterogeneous constructionism, researchers are always already harnessing diverse resources as they seek to establish knowledge. Could something simpler than a map of their complex situatedness move them to mobilize new resources and organize them to support new directions in their work? Suppose, in this spirit, I let the complexity of situatedness stay in the background and focused on helping researchers to conceptualize directions that would address more complexity in the situations they studied. If I were effective in shifting the situations that interest them, I could hope that they would try to identify and address for themselves the practical considerations associated with making changes to their research. Yet what would crystallize for researchers a decision not "simply to continue along previous lines"? Could something simpler than maps guide them as they mobilized different resources or organized them in new directions?

Two terms emerging from the explorations described in previous chapters helped keep me oriented, so I began to bring them to the attention of others through my teaching and presentations. Let me spell out their meaning and implications.

Unruly Complexity

I began to use the term *unruly complexity* after I reconstructed Picardi's modeling (chapter 4) and reflected on this process (section A of this chapter). The set of contrasts I had used to probe Picardi's modeling (table 4.1), the components of heterogeneous construction (end of chapter 4), and the challenges involved in web analysis (see section A of this chapter) could be synthesized into one overall contrast; namely, between conceptual and practical commitments that *render a complex situation system-like* (fig. 5.4) and moves that help researchers *represent and engage with unruly complexity* (fig. 5.5). Let me summarize this contrast in words:

When researchers assume that there are systems with clearly defined boundaries, coherent internal dynamics, and simply mediated relations with their external context, they can locate themselves outside those systems and seek generalizations and principles affording a natural or economical reduction of complexity. A contrasting image is that well-bounded systems, when they are encountered, require explanation as special cases of unruly complexity, in which boundaries and categories are problematic, levels and scales are not clearly separable, structures are subject to restructuring, and components undergo ongoing differentiation in relation to one another. Control and generalization are difficult, and no privileged standpoint exists; ongoing assessment requires engagement in the situation.

The system versus unruly complexity contrast can be employed at a number of levels in ecology and in interpretation of science:

1. It can be used as a checklist of the ways in which any particular analysis might promote a system-like approach to complexity or grapple with unruly complexity. Researchers could then consider the consequences for their analyses of replacing any system-like item with the corresponding alternative, or vice versa.

2. It can serve as an overarching name that could be adopted by researchers dissatisfied with some or all of the assumptions that render complexity

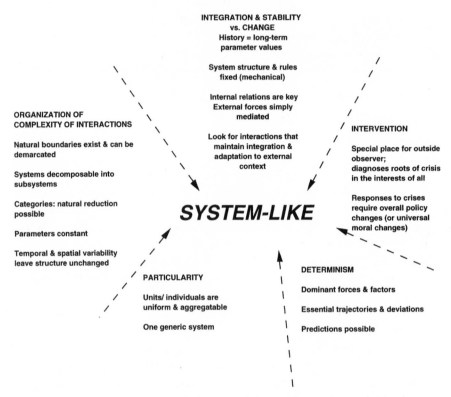

Figure 5.4 Conceptual and practical commitments that help researchers render complex situations into well-bounded systems.

system-like. Having a name that crystallizes for those researchers the directions in which they want to push their work might motivate and orient their subsequent efforts.

3. It can serve as a checklist that researchers use to reflect on what it would mean *practically* to pursue any of the alternatives. Researchers could follow the method I used to interpret Picardi's work (chapter 4, section B) to expose, in a somewhat systematic manner, the heterogeneous resources used in the construction of any particular analysis and to identify potential sites of engagement for mobilizing alternative resources.

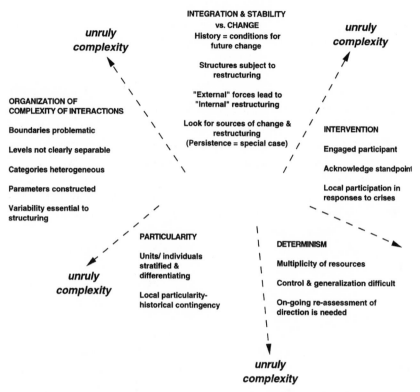

Figure 5.5 Conceptual and practical moves that help researchers represent and engage with the unruliness of complexity

Intersecting Processes

Intersecting processes is the other term I have used to help students and researchers conceptualize directions that would address more complexity in socio-environmental studies (Taylor and García-Barrios 1995; Taylor 2001a). The term overlaps conceptually with *unruly complexity*—to analyze social and environmental change as something produced by *intersecting* economic, social, and ecological processes that *operate at different scales* requires attention to the ways in which these processes are unruly; that is, transgress boundaries and restructure "internal" dynamics, thus ensuring that socio-environmental situations do not have clearly defined boundaries and are not simply governed by coherent, internally driven dynamics. However, I use the term *intersecting processes* to suggest that different strands of the processes can be teased out in a somewhat disciplined fashion. This will be evident in the following presentation of a case of soil erosion in a mountainous agricultural region in Oaxaca, Mexico, which I have based on the analysis of Mexican colleagues Raúl García-Barrios and his brother Luis (García-Barrios and García-Barrios 1990).

The severe soil erosion evident now in the municipality of San Andrés is not the first occurrence of such a problem in the region. After the Spanish conquest, when the indigenous population collapsed from disease, the communities abandoned their terraced lands, which then eroded. The remaining populations moved to the valleys and adopted laborsaving practices from the Spanish, such as cultivating wheat and using plows. As the population recovered during the eighteenth and nineteenth centuries, collective institutions evolved that reestablished terraces. Erosion was reduced, soil dynamics were stabilized, and perhaps some soil accumulation was stimulated. But this type of landscape transformation needed continuous and proper maintenance. If a terrace were allowed to erode, the soil would wash down and damage lower terraces; there was the potential for severe slope instability.

What made the necessary maintenance possible were collective institutions, which first revolved around the Catholic Church and then, after independence from Spain, around rich Indians called *caciques*. These institutions mobilized peasant labor for key activities—not only maintaining terraces, but also sowing maize (corn) in work teams and maintaining a diversity of maize varieties and cultivation techniques. The caciques benefited from what was produced, but were expected to look after the peasants in hard times—a so-called moral economy (Scott 1976). Given that the peasants felt security in proportion to the wealth and prestige of their cacique, and given that prestige attached directly to each person's role in the collective labor, the labor tended to be very efficient. In addition, peasants were kept indebted to caciques and could not readily break their unequal relationship. The caciques insulated this relationship from change by resisting potential laborsaving technologies and ties to outside markets in maize.

The Mexican Revolution ruptured this closed system of reciprocal obligations and benefits by taking away the power of the caciques and opening the communities to the changing outside world. Many peasants migrated to industrial areas, sending cash back or bringing it with them when they returned to their communities. Rural population declined, transactions became monetarized, and prestige no longer derived from one's place in the collective labor. With monetarization and loss of labor, the collective institutions collapsed, and terraces began to erode. National food-pricing policies favored urban consumers, which meant that maize was grown only for subsistence needs in rural communities. Little incentive remained for intensive agricultural production. New laborsaving activities, such as goat herding, which contributes in its own way to erosion, were taken up without new local institutions to regulate them.

Figure 5.6 is a diagram I drew to help me narrate this story to others and to highlight a number of themes pertaining to intersecting processes. The following discussion illustrates how socio-environmental studies like this one, especially studies of political ecology (Peet and Watts 1996a), can provide rich material for exploring the problematic boundedness of ecological complexity (one of the shifts of emphasis mentioned at the end of chapter 4) and for amplifying the themes introduced in earlier chapters.

1. **Intersecting processes involve inseparable dynamics.** Processes of different kinds and scales, involving heterogeneous elements, are interlinked in the production of any outcome and in their own ongoing transformation. Each is implicated in the others (even by exclusion, as when caciques kept maize production during the nineteenth century insulated from external markets). Notice especially the relationship between environmental degradation and the population *decline* shown in the top strand. This association can be used to grab the attention of environmentalists who identify population *growth* as a major environmental issue. However, it is neither population decline nor growth, but labor, that was important in this case. Labor is something defined by the technologies of production (the second strand) and the social institutions that govern it. Such institutions operate both locally (the third strand) and at places distant from where the erosion occurs (the fourth strand). In short, the relationship between population and environmental change was highly mediated, depending on the technologies used and the local and national social and economic institutions through which labor and production were organized. No one kind of thing, no single strand on its own, is sufficient to explain the currently eroded hillsides. (This theme can be extended to call into question other explanations for environmental degradation that center on a single dynamic or process, for example, climate change in erosive

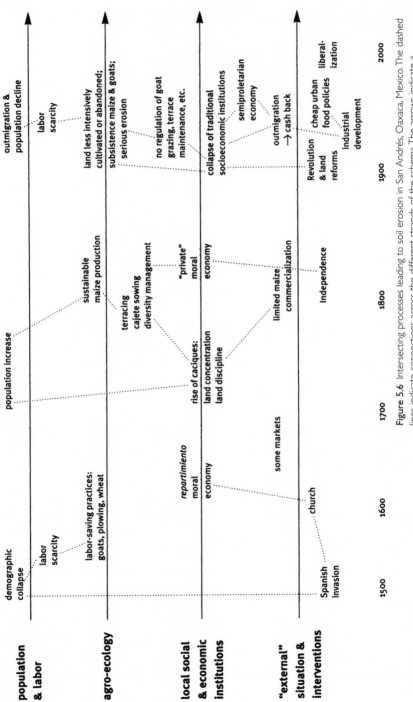

Figure 5.6 Intersecting processes leading to soil erosion in San Andrés, Oaxaca, Mexico. The dashed lines indicate connections across the different strands of the schema. The arrows indicate a continuation of the element over time. (From Taylor 1997a.)

population & labor

1500 demographic collapse / labor scarcity

labor-saving practices: goats, plowing, wheat

population increase

sustainable maize production

outmigration & population decline

labor scarcity

agro-ecology

terracing cajete sowing diversity management

land less intensively cultivated or abandoned; subsistence maize & goats; serious erosion

no regulation of goat grazing, terrace maintenance, etc.

local social & economic institutions

repartimiento moral economy

rise of caciques: land concentration land discipline

"private" moral economy

collapse of traditional socioeconomic institutions

semiproletarian economy

"external" situation & interventions

Spanish invasion

church

some markets

limited maize commercialization

Independence

Revolution & land reforms

industrial development

outmigration → cash back

cheap urban food policies

liberal-ization

1500   1600   1700   1800   1900   2000

landscapes, increasing capitalist exploitation of natural resources, or modernization of production methods.)

The theme of inseparable dynamics can be teased out into four aspects:

2. **In intersecting socio-environmental processes,** *differentiation among unequal agents is implicated.* Sustainable maize production depended on a moral economy of cacique and peasants, and the inequality among those agents resulted from a long process of social and economic differentiation. Similarly, the demise of this agroecology involved the unequal power of the state over local caciques, of urban industrialists over rural interests, and of workers who remitted cash to their communities over those who continued agricultural labor.

3. **Heterogeneous elements and scales are involved.** The situation has involved processes operating at different spatial and temporal scales, encompassing elements as diverse as the local climate and geomorphology, social norms, work relations, and national political economic policy.

4. **Historical contingency is significant.** The role of the Mexican Revolution in the collapse of nineteenth-century agroecology reveals the contingency that is characteristic of history. The significance of such contingency rests not on the event of the revolution itself, but on the different processes, each having a history, with which the revolution intersected.

5. **Structuredness is not reducible to micro- or macro-determinations.** Although there is no reduction to macro- or structural determination in the account of soil erosion, the focus is neither on local, individual-individual transactions nor on the complex patterns produced by multiple simple transactions. Regularities, such as the terraces and the moral economy, persist long enough for agents to recognize or abide by them. That is, structuredness is discernible in the intersecting processes.

My synopsis and diagram of the García-Barrioses' more detailed account have a number of implications for thinking about the agency of the people studied and, reflexively, that of researchers reconstructing intersecting processes:

6. **The account represents** *agency as distributed across different kinds of agents and scales,* **not something centered in one class or place (Thompson 2002).** In the nineteenth-century moral economy, caciques exploited

peasants, but in a relationship of reciprocal norms and obligations. Moreover, the local moral economy was not autonomous; the national political economy was implicated, by its exclusion, in the actions of the caciques that maintained labor-intensive and self-sufficient production. Although the Mexican Revolution initiated the breakdown in the moral economy, the ensuing process involved not only political and economic change from above, but also from below and between—semi-proletarian peasants brought their money back to the rural community and reshaped its transactions, institutions, and social psychology.

7. The account has an *intermediate complexity*—it is neither highly reduced nor overwhelmingly detailed. The elements included in my synopsis and in the diagram are heterogeneous, but I tease out different strands. The strands, however, are cross-linked; they are not torn apart. By acknowledging this intermediate level of complexity, the account steps away from debates centered on simple oppositions, such as ecology-geomorphology versus economy-society or ecological rationality versus economic rationality. Similarly, by placing explanatory focus on the ongoing, intersecting processes, the account discounts the grand discontinuities and transitions that are often invoked, such as peasant to capitalist agriculture or feudalism to industrialism to Fordism to flexible specialization.[23]

8. Intermediate complexity accounts *favor the idea of multiple, smaller engagements linked together within the intersecting processes.* My synopsis and diagram can be read as an engagement with current scholarly discourse in an effort to promote the concept of distributed agency. This concept has implications not only for how environmental degradation is conceptualized, but also for how one responds to it in practice. Intersecting processes accounts do not support government or social movement policies based on simple themes, such as economic modernization by market liberalization, sustainable development through promotion of traditional agricultural practices, or mass mobilization to overthrow capitalism.

9. This shift in how policy is conceived suggests a corresponding shift in scholarly practice: at the level of research organization, intersecting processes accounts *highlight the need for transdisciplinary work grounded in particular locations.* They do not underwrite the customary multidisciplinary projects directed by natural scientists, nor economic

analyses based on the kinds of statistical data available in published censuses.

10. Finally, intermediate complexity accounts *preserve a role for some kind of social scientific generalization.* The synopsis and diagram abstract away an enormous amount of detail, a move that suggests that the particular case described by the García-Barrioses might be relevant to other cases. The account does not provide a general explanatory schema, but it could at least serve as a template to guide further studies. Such a template would be elaborated in new research projects once researchers began to address the particularities of the situation they were studying. In other words, the particularities of each case would not warrant starting from scratch when attempting to understand and engage in socio-environmental change. The intermediate complexity of my account also means—and here I am applying some reflexivity to my own representational work—that I have deflected attention away from the need to examine the particular institutional and personal re-sources, agendas, and alliances that people like me would have to culti-vate to gain support for the desired transdisciplinary research or policy interventions.

▲

In the course of working to interest more scientists and interpreters of science in analyzing scien-tists' diverse resources and paying attention to their distributed agency, I had formulated two fur-ther themes about interpreting science:

   1. When interpreters of science deal with scientists' webs of heterogeneous resources— even if only to discount those webs' complexity—they must be building their own webs. In other words, all research involves heterogeneous construction (section A).

   2. Those who interpret research as heterogeneous construction should try to distribute to others the work of interpreting and engaging with that research (section B). That is, they should lessen the pressure on themselves or on any one person to convey the full com-plexity of the researchers' resources. A single individual should not even be relied on to deliver the resources needed for others to expose this complexity. Indeed, when I intro-duced the terms unruly complexity and intersecting processes to students and col-leagues (section C), I hoped that this would help them to conceptualize directions that would address more complexity in the situations they studied, but I relied on them to take initiative in mobilizing new resources and organizing them to support new directions in their work.

   A tension had become apparent. In recognition of the heterogeneous construction both of science and of its interpretation, I was working to stimulate others to identify the diverse resources *mobilized by particular agents who span different domains of social action. As I did so, however, I made conceptual and methodological choices that, to varying degrees, pushed the complexity of my own and my audience's sociality into the background. In principle, the practical conditions behind the interpretive choices made by researchers such as myself can always be opened up for reconstruction. The complexity that has been hidden can be brought back into the foreground. But when, in practice, is practical reflexivity worth pursuing? This remained a matter for further investigation and experimentation …*

▲

6

Reasoned Understandings and Social Change in Research on Common Resources: Introducing a Framework to Keep Tensions Active, Productive, and Ever-Present

As a conceptual matter, I had not finished exploring practical reflexivity in relation to the heterogeneous construction of research and its potential reconstruction. Yet, as a practical matter, it was necessary to reach audiences comfortable with the convention of presenting scientific accounts or interpretations as if they could stand independently of the author's and the audience's particular situatedness. As I continued to wrestle with this tension, my teaching of socio-environmental studies and interpretation of science suggested a lead worth following.

In most of my interdisciplinary classes, students lacked the sustained research experience that could be shared in mapping workshops, but they were also usually free of commitments to any specialized area of ecological or interpretive research. This combination of constraint and opportunity led me to formulate themes that I could introduce through cases accessible to a wide range of students, which, although simple to convey, would point to the greater complexity of particular cases and to further work needed to study them. I described these as opening-up *themes.*

It seemed that the same basic approach might be tried out on nonstudent audiences in which no one area of specialization predominated. The idea was to formulate themes that stimulated members of the audience to examine the particularity in practice of their own contributions to changing knowledge, society, and ecology. If this approach was effective, I could afford to push situatedness into the background of my presentations without abandoning my perspectives on heterogeneous construction and on representing-engaging with ecological complexity. Another way of expressing this challenge was that I wanted to acknowledge the tension between, on one hand, the multiplicity of particular situated complexities of my audience's knowledge-making and, on the other hand, the simplicity and apparent generality of the themes. My aim was to keep that tension active, productive, and ever-present. To this end, I developed the multipart framework that I introduce in this final chapter, which I illustrate using case material from research on people who manage natural resources that are held in common. (See also Taylor 2001d for an earlier application in the context of population-environment research.) Along the way, I draw together many of the themes and some of the cases of the earlier chapters ...

◆ ◆ ◆

A. Researchers Conduct a Dialogue, Involving Concepts and Evidence, with the Situations Studied

What can agents do (when resources are held in common)?

Garrett Hardin's (1968) idea of the "tragedy of the commons" is widely invoked in discussions of conservation and natural resource management (e.g., Picardi and Seifert 1976; see chapter 4, section B). In his hypothetical common pasture, each herder in the community follows the same logic: "I will receive the benefit in the short run from increasing my herd by one animal; everyone will share any eventual cost of diminished pasture per animal; therefore I will add another animal to my herd." Overstocking and pasture degradation is thus inevitable. The same model has been applied to a range of environmental and social resources, from CO_2 emissions to library books (Berkes et al. 1989).

Many teachers of environmental studies use classroom simulations to introduce students to the "tragedy of the commons" and its implications. I have designed an activity in which students are asked to take the role of herders who are all given the same initial amount of cattle and cash. Each year they have an opportunity to buy cows to add to their herd, and they receive income from the sale of milk and excess calves. I sum up everyone's purchases and use a formula to calculate a figure for the income per cow during the year from milk and calves. In this formula, the income declines once the combined herd on the common pasture exceeds some threshold and the pasture becomes overgrazed. I inform the herders of the income per cow, and each of them updates their cow and cash tally. The only other stipulation is that, on my own, I make no more rules. Herders have to decide whether they want additional rules and how to get them implemented in their community.[24]

Before reading further, ask yourself what purchasing strategy you would use if participating in this simulation and what rules you would try to get implemented. Try this even if you are familiar with the idea of the tragedy. Because some of you may be tempted to skip ahead to look for "the answer the teacher wants," I will break the text at this point with a photo from the West African country of Mali of actual herders who use a common rangeland.

Figure 6.1 A photo of actual herders who use a common rangeland in the West African country of Mali. (Courtesy of Matthew Turner.)

OK, readers—what did you come up with? An obvious response is private ownership of the land—or whatever resource is used in common—so that individuals have to factor in the full costs of their decisions. Another remedy is government control to "restrain people who find it irrational to restrain themselves" (McCay 1992, 189). Hardin claimed that, unless resources are privatized or there is government coercion, individual self-interest leads inevitably to the overexploitation and degradation of resources held in common. He also invoked evolutionary arguments to forestall any questioning of the premise of self-interested agents (see also Picardi and Seifert 1976). That is, if there happened at one time to be some individuals who restrained themselves from increasing their herds, they would have fewer resources than individuals who did not, and so would be buffered less in bad times or have fewer surviving offspring. Sooner or later, this kind of individual and their restraint would go extinct, and the tragedy would unfold.

Some of you, in contrast to Hardin, might have proposed taking turns using the pasture, with the length of each herder's stint determined by someone appointed by the community to monitor the state of the pasture. To this proposal you might have added sanctions against over-stinting, to be enforced by the community as a whole or its authorized representatives. If so, you pointed in the direction taken by a growing body of socio-environmental research on the management of actual non-privatized common resources since Hardin's essay. Many cases have been documented in which people, communicating and working together in communities, successfully build and maintain local institutions for managing a resource held in common (see Berkes et al. 1989; Ostrom 1990; McCay and Jentoft 1998).

The lessons emerging from the "post-Hardin" research are not, however, the focus of the class. Instead, my goal is to show how the simulation can be extended so that students appreciate shortcomings in the "tragedy" model and consider the ways in which people analyze ecological and social complexity more generally. The simple, but very influential, "tragedy of the commons" model provides rich material for critical thinking about the dynamics of environmental change.

Four Levels in Students' Responses

Recall my stipulation that student-herders have to work out for themselves whether they want additional rules and how to get them implemented in their community. In my experience, students begin to express objections as the simulation progresses, and some attempt to mobilize fellow herders into adding or changing rules. Usually the responses do not gel in time to prevent dire overgrazing, and the herders' annual income drops almost to zero. I then call time out and review what has happened. First, the group's combined income is much less than it could be. Second, the group has differentiated—the initial equality among herders has given way to large disparities in wealth. I ask students to keep these observations in mind and negotiate what to do now that the overgrazing has occurred. In the lively discussions that follow the review, certain voices count more than others. Herders who have the largest herds and greatest wealth can use their resources to exert disproportionate influence, not only on what propositions are accepted, but also on the procedures for making decisions. Students who had purchased few or no cattle because they did not want to contribute to overgrazing are poor and have less influence. (Readers are welcome to take a break at this point to formulate the negotiating strategy you might use at this point in the simulation.)

The changes that the students seek during the initial phase of the simulation and after the review typically fall into four levels. (Notice where your initial approach to the simulation, as well as the negotiating strategy you just formulated, fits.) The students

1. want more realism or detail in the rules—to allow cattle to die, purchase prices to vary, herders to trade among themselves, income to vary with season, and so on. They seek such changes even though the changes do not prevent overgrazing.

2. communicate about their actions, plans, and norms (e.g., "greedy herders should be shunned").

3. allow exchanges with the outside world. For example, the simulation assumes that cattle can be bought from some unspecified place, and that milk and calves can be sold. Cattle themselves, then, ought to be saleable. Some students even propose to leave the game to become agriculturalists, traders, or urban workers.

4. get involved in the politics of collective governance; that is, in conflicts and negotiation among unequal parties. Typical proposals are to halve every herd, set a common upper limit on all herds, tax large herds, and privatize pasture. To institute any proposal, however, turns out to be more difficult. Poor, conservation-minded herders see the halving proposal as unfair to them, while wealthy herders tend to use their muscle to resist leveling proposals. If land is privatized, for example, the wealthy want it to be subdivided in proportion to current unequal herd sizes. Many students, when faced with the stratification of wealth and influence, want to begin again with the initial conditions of equality. As the teacher, I insist it is too late for that—after all, that state exists nowhere in the known world. Some students then try to invoke an outside government (see level 3) with power to impose such changes over the objections of the wealthy herders.

Through their responses, my students communicate with one another, make exchanges with the outside world, and negotiate conflict and cooperation among unequal parties. Broadly speaking, these aspects of social life match post-Hardin research on actual common resources. However, as I mentioned earlier, the class does not focus on the lessons of that research.[25] Instead, I explore with students the challenge their responses pose to the fundamental assumption that the world is composed of systems in a strong sense of natural units that have clearly defined boundaries and coherent, internally driven dynamics.

Each of the levels at which students seek change can be rephrased in themes that disturb Hardin's idea and, more generally, simple models of well-bounded systems:

1. Instead of viewing the system as composed of individuals whose interactions are given at the outset, allow the system's dynamics to be mutable.
2. By thinking about the networks of social support in which "individuals" are raised and in which they then operate as adults, take "sociality," not individuality, to be primary. Atomized individuals in the form Hardin presented are unlikely when networks of social support make communication—even through people's silence—unavoidable. Such networks give power to sanctions, in the form of withdrawal of social links, and thus also strengthen the threat of such sanctions.
3. View any boundaries defining a system as permeable.
4. Consider how inequality among individuals within the system colors the paths individuals can pursue, including their responses to developments "outside" the system (e.g., when wealthy herders accept government-imposed privatization but manage to ensure that land is allocated in proportion to current herd sizes).

A corollary of these themes is the need to be alert to any situation in which individuals appear atomized and noncommunicating and to inquire into the history that led up to that Hardin-like state of affairs. In other words, the corollary invites investigation of the special conditions that make a simple model applicable. (This corollary recalls the conclusion of chapter 1, section B, that there is hidden complexity in apparently simple models.)

The four themes and the corollary open up an alternative to the perspective that environmental change can be understood in terms of natural units that have clearly defined boundaries and coherent, internally driven dynamics. These themes are simple enough to convey to students, but point to the need for further work to address the complexity of particular cases. As a further stimulus to moving in that direction, I provide students with a glimpse of the kinds of considerations that might be involved in analyzing the complexity of politics, sociality, and environmental change in a particular commons. The following brief summary of research during the 1980s on nomadic pastoralists serves this purpose (see also chapter 4, section B).

Differentiated Agents Situated in Intersecting Processes

Nomadic pastoralists are herders living in semiarid climates where rainfall is variable, unpredictable, and spatially patchy, who spend at least part of their year roaming a common rangeland in search of patches of pasture. During the 1970s, the dominant accounts of pastoralism had an environmental determinist outlook, in which range degradation and desertification were attributed to pastoralists who allowed grazing beyond the environment's supposed carrying capacity. Problems in this picture were exposed by research during the 1980s on the ongoing transformations of the economies and ecologies of nomadic pastoralist groups (chapter 4, section B; see also Taylor and García-Barrios 1995). The alternative picture highlights different factors that are implicated—to different degrees in different locations—in past transformations, such as taxation, establishment of military control, imposition of borders, and other aspects of colonial and postcolonial administration. Similar kinds of factors are implicated in more recent changes: further severe droughts and extension of agricultural areas; regulation of conflict over resources; privatization of access to resources; sedentarization and other development projects sponsored by national governments and international agencies; and the changing economic conditions and terms of trade accompanying structural adjustment.

Some pastoralist societies have been rapidly restructuring so that their boundaries have become permeable. Pastoralists break their reciprocal relations with agriculturalists to become cultivators themselves; better-off agriculturalists become absentee herd owners, and the poorer peasants and herders become their hired laborers. Squeezed for time to take their own herds out on the better rangeland, these herder-laborers allow their livestock to overgraze areas close to their settlements. As a result, environmental degradation, where apparent, lies close to population concentrations, not out on the common rangeland (Little 1985, 1987; chapter 4, section B).

This alternative picture of nomadic pastoralism exemplifies the four aspects of the system-disturbing perspective (themes 1–4 of the previous section) and suggests some extensions:

- *Structures are subject to restructuring* (e.g., nomadic pastoralism becomes combined with and constrained by agricultural activities). "External" influences are usually implicated in the structuring of any "local" institutions (see themes 1 and 4).
- *Categories and the boundaries between them are problematic* (e.g., pastoralism/agriculture; herding/laboring; climatic/economic forces). Levels and

scales are not clearly separable (e.g., local, national, and international practices, knowledge, and policies all enter the dynamics of the pastoral situation)
(see themes 2, 3, and 4).

- *Control and generalization are difficult.* For example, there may be plenty of
  degraded common property resources worldwide. However, general policy
  solutions are not warranted once such situations are seen as transformations
  of existing complex and differentiated politics, not as the inevitable result of
  some fundamental, apolitical dynamic (see themes 2 and 4). This caution
  applies equally to government-imposed privatization and to NGO promotion
  of traditional pastoral and agricultural practices. The alternative picture,
  instead of pointing to policies based on simple themes, exposes multiple
  possible engagements by various agents in a range of social positions.
  Moreover, this multiplicity opens up questions about the ways in which any
  particular engagement, when linked with others, might contribute to desired
  and unintended restructurings.

The term *unruly complexity* encapsulates the preceding aspects of
the system-disturbing perspective. I use the allied term *intersecting
processes* to convey the idea that social and environmental change can
be analyzed as something produced by intersecting economic, social,
and ecological *processes* that operate at different scales (fig. 6.2; Peet and
Watts 1996a). These processes transgress boundaries and restructure
"internal" dynamics, thus ensuring that socio-environmental situations
do not have clearly defined boundaries and are not simply governed by
coherent, internally driven dynamics. (For elaboration, see chapter 5,
section C.)

Three Formulations of Complexity and Three Angles from Which to View the Practice of Researchers

Let me reflect on the structure of this tragedy of the commons class as a
whole. The simulation begins with a simple, well-bounded system, in
which equal agents act independently in their own self-interest according
to Hardin's well-known theory. In contrast, the lesser-known work summarized in the last section analyzes dynamics among particular, unequal
agents whose actions implicate or span a range of social domains. I did not,
however, go so far as to present any one particular case in detail, examine
the dynamics among particular unequal and differentiating agents, and
expose their locally centered but scale-spanning character. In general, such
dynamics are difficult to convey in a standard-length lecture and difficult

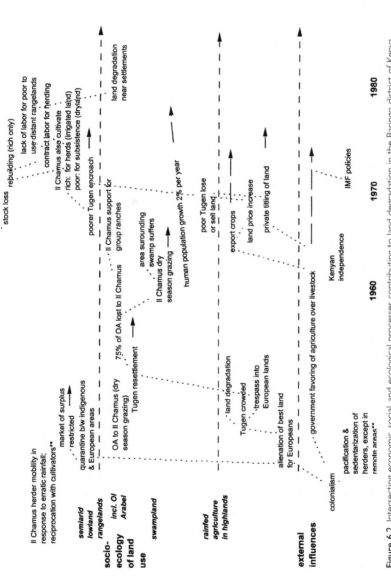

Figure 6.2 Intersecting economic, social, and ecological processes contributing to land degradation in the Baringo district of Kenya. Following the same conventions as figure 5.6, the dotted lines indicate connections across the different strands of the schema. The asterisks also denote two connected elements. The arrows indicate a continuation of the element over time. See also discussion of nomadic pastoralism in this chapter and in chapter 4, section B, contrasts 4 and 6. (Adapted from Little 1987.)

The following labels appear in the figure:

severe droughts

stock loss
rebuilding (rich only)

lack of labor for poor to
use distant rangelands

contract labor for herding

Il Chamus also cultivate
rich: for herds (irrigated land)
poor: for subsistence (dryland)

land degradation
near settlements

market of surplus
restricted

quarantine b/w indigenous
& European areas

Il Chamus herder mobility in
response to erratic rainfall;
reciprocation with cultivators**

poorer Tugen encroach

OA to Il Chamus (dry
season grazing)

Tugen resettlement

75% of OA lost to Il Chamus

Il Chamus support for
group ranches

area surrounding
swamp suffers

Il Chamus dry
season grazing

human population growth 2% per year

**semiarid
lowland
rangelands**   *incl. Ol
Arabel*

**socio-
ecology
of land
use**

swampland

land degradation

Tugen crowded

trespass into
European lands

alienation of best land
for Europeans

poor Tugen lose
or sell land

export crops

land price increase

private titling of land

**rainfed
agriculture
in highlands**

government favoring of agriculture over livestock

**external
influences**

colonialism

pacification &
sedentarization of
herders, except in
remote areas**

Kenyan
independence

IMF policies

1960   **1970**   **1980**

for members of an audience to digest and make their own. The system-like formulation of the tragedy, in contrast, is simple enough to be readily communicated.

The classroom simulation was designed precisely to acknowledge and preserve the tension between particular complexity that is difficult to convey and the (apparent) simplicity of system-like formulations. The simulation was designed to uncover themes that are easy to absorb and adapt, yet, at the same time, open up issues, pointing to greater complexity and to further work needed in particular cases. Unlike system-like formulations, these themes build in a reminder of their limits: they tell us that further work in particular cases is needed—which may qualitatively change the analysis—but they do not demonstrate how that work is to be done. In summary, the structure of the tragedy of the commons class involves inserting this type of formulation between two others in a way that preserves the tension between them:

Type 1. Simple formulations of well-bounded or strong systems (e.g., Hardin's tragedy of the commons)

Type 2. Simple themes that open up issues, pointing to greater complexity and to further work needed in particular cases (e.g., the classroom simulation)

Type 3. Work based on dynamics that develop over time among particular, unequal agents whose actions implicate or span a range of social domains (e.g., research from the 1980s on nomadic pastoralism, summarized above)

Insertion of the opening-up themes constitutes an opening-up theme in its own right. Up to this point, the chapter's discussion of commons research has viewed the practices of researchers in the traditional way; namely, as a *dialogue* conducted with the situations they study, employing concepts and evidence to produce their accounts—in this case, accounts of common resources. Yet, as soon as we pay attention to how readily these accounts can be conveyed and digested, we have moved away from a simple view of knowledge-making as a dialogue with the situation studied. We have begun to view the practices of researchers from an additional angle; namely, through the interactions with other social agents through which researchers endeavor to establish what counts as knowledge. Researchers always have to engage others in accepting and applying their accounts; they cannot establish knowledge on their own. When an audience questions or modifies their accounts, a dialogue among knowledge-makers ensues. This angle on research adds complexity to the traditional view; further work is needed to make sense of the two simultaneous dialogues— with the situation studied and with audiences that include knowledge-users and other knowledge-makers.

This last paragraph has distinguished two angles from which scientific practice can be viewed:

A. Researchers conduct a dialogue, involving concepts and evidence, with the situations studied.
B. Researchers endeavor, through interactions with other social agents, to establish what counts as knowledge.

Later in this chapter, I use another opening-up theme to introduce a third angle:

C. Researchers pursue social change by self-consciously addressing the complexities of their own social situatedness as well as the complexities of the situations they study.

The three angles, together with the three types of formulations of complexity, define nine combinations (A1–C3) that will be discussed in this chapter (fig. 6.3). This 3 × 3 framework captures the implications of intersecting processes, heterogeneous constructionism, and practical reflexivity. I can use this framework when lecturing or writing to audiences who do not share one area of specialization. The class on the tragedy of the commons encompasses the three types of formulations from angle A. The reflection of the last two pages, serving as another type 2 formulation, opens up angle B, reminding us that social interactions to establish knowledge are always implicated in angle A. However, before exploring angle B and, later, angle C, I want to present an alternative opening to angle A, one directed at commons researchers. This presentation will prefigure the addition of angle C and allow me connect this chapter as a whole more firmly to the discussions of agency in previous chapters.

What can agents do—What social changes can researchers facilitate with their various understandings?

Let me begin as before: In Garrett Hardin's hypothetical common pasture, each herder in the community follows the same logic: "I will receive the

	The practice of researchers can be viewed from different angles, which highlight the researchers'...		
The resulting formulations involve...	A. dialogue with the situation studied	B. social interactions to establish what counts as knowledge	C. efforts to pursue social change in which they address self-consciously the complexities of their own situatedness as well as of the situation studied
1. simple, well-bounded systems (coherent internal dynamics and simply mediated relations with their external context)	e.g., Hardin's model in which atomized, self-interested individuals produce the tragedy of the commons	e.g., The content of scientific knowledge is determined by the social aspects of knowledge production. (Alternatively: sociality has at most a transient effect.)	e.g., Researchers preserve terms familiar to their audience and thus leave most aspects of their social situatedness taken for granted.
2. simple themes that point to greater complexity and further work needed in particular cases	e.g., Classroom simulation of commons management that adds politics and communication among unequal agents and exchanges with the outside.	e.g., Rhetorical interpretations of tragic and locally managed commons, which direct attention to the actions and agents that follow from accounts of the commons.	e.g., Mapping workshops, in which self-selected scientist-participants expose and explore their social context of research and their research at the same time.
3. work based on dynamics that develop over time among particular, unequal agents whose actions implicate or span a range of social domains	e.g., Case of changes in pastoralism and agriculture in Kenya as analyzed by P. Little.	e.g., Web of heterogeneous resources involved in the construction of Picardi's computer modeling of nomadic pastoralists.	? (participatory restructuring of knowledge making and social change by diverse agents)

Figure 6.3 Three angles from which to view a researchers' practice (A, B, C) and three types of formulations of complexity (1, 2, 3), with an example drawn from this chapter for each of the nine combinations (A1–C3).

benefit in the short run from increasing my herd by one animal; everyone will share any eventual cost of diminished pasture per animal; therefore I will add another animal to my herd." Overstocking and pasture degradation is thus inevitable.

Now let me diverge from the earlier opening: Since the mid-1980s, the institutions through which non-privatized, common resources are managed have been examined by a growing number of socio-environmental researchers, many of them coming together under the umbrella of the International Association for the Study of Common Property. Actual agents, it appears, often do better than those in Hardin's thought experiment. People, working together in communities, overcome their short-term self-interest and build local institutions for managing a resource held in common (Berkes et al. 1989; McCay and Jentoft 1998). In general, successful institutions are operated and monitored by a clearly defined group of those directly concerned with the resource. They are "externally accepted"—that is, the government, markets, or industries tolerate or even support jurisdiction over the resource by the community of users (Ostrom 1990).

How are we to make sense of the contrast between the "tragic" and the locally managed, externally accepted commons? The answer depends on which "we" one identifies with. Many kinds of agents are, in different ways, involved in the commons. There are socio-environmental researchers who provide accounts of successful resource and environmental management, and a range of other writers who still invoke Hardin's idea. There is the local community of users, as well as agents of the government, commerce, and industry—agents who either restrain outsiders from exploiting the resource in question or who are themselves among those restrained. Finally, there are some observers and analysts of scientific change who try to make sense of discourse on the commons since Hardin's initial contribution.

Let me enter this arena by focusing on the socio-environmental researchers and pick up on the central concern of the commons discourse; namely, that the actions of rational agents can contribute to undesired collective, social outcomes. In this context, it is pertinent to ask commons researchers to consider the relation between their own reasoned understandings and the effects on society that result. That is, we can ask a question that is complementary to "What can agents do?" namely, "What social changes can researchers facilitate with their various understandings?" The relevant understandings concern not only the situations that researchers study, but also the social situations in which those researchers are embedded. The "social change" may be as modest or local as stimulating members of the audience to change the concepts they use, or as ambitious as stemming the resource degradation in some environmental situation. Yet, at all points on this spectrum, the linkages between understanding and agency invite systematic examination.

The first step in this examination is to consider the practice of researchers from the angle of their dialogue with the situations they study (angle A)—how they employ concepts and evidence in producing their accounts of common resources. One formulation of this dialogue involves simple formulations of well-bounded systems.

A1. The Tragic and the Locally Managed Commons: Two System-like
Formulations

The dynamics of Hardin's "tragic commons" are defined in terms of equal
individuals who function in the same self-interested way. Unlike these
a priori hypothetical dynamics, the "locally managed commons" stems from
a wealth of studies of actual institutions that attempt to manage resources
held in common (Ostrom 1990; Feeny et al. 1990). Ostrom (1993) summarized the conditions for the success of those institutions in a set of design
principles (see also McKean and Ostrom 1995):

1. Clearly defined boundaries of the resource, and of the community of users.
2. Benefits of resource use proportional to the costs imposed for its
   maintenance and management.
3. Users affected by rules of resource use are involved in deciding on any
   changes to those rules.
4. Infractions of rules are monitored; monitors are users or are accountable
   to them.
5. Sanctions are graduated according to the severity of the offense.
6. Conflict resolution mechanisms are rapid, low-cost, and local.
7. External authorities and other interested persons accept jurisdiction
   over the resource by the resource users' institutions.
8. Institutions for managing large resources form nested layers of
   organization.

The formulation of the locally managed commons differs dramatically
from the unregulated, open-access character of the tragic commons
in emphasizing institutions of collective governance. However, it has
not given much attention to inequality, permeable boundaries, or the
processes whereby institutions of collective governance evolve. In both
the locally managed and the tragic commons, self-interest is the basis of
people's rational actions (their "rational choices"), and the social and
environmental phenomena of interest are system-like. That is, a clear
boundary is drawn between what goes on inside the system and what
influences it from the outside. The outside influences set the parameters
within which the system operates, but the focus is on the dynamics inside
the boundary.

With this commonality as a backdrop, the second formulation (A2)
would encompass the previous description of the class simulation used to
introduce a broad non-system perspective on the dynamics of environmen-

tal change. As before, this would lead into the third formulation (A3), consisting of specific research on differentiated agents situated in intersecting processes. The earlier reflection on the class simulation could follow the three formulations (A1–A3) and open up inquiry into the interactions researchers have with other social agents when they are working to establish what counts as knowledge. This should reveal that angle B is always implicated in angle A.

There is a complementary opening-up theme that follows from comparing the actions or policies that correspond to the different types of formulations and the kinds of agents involved.[26] The dynamics of the tragic commons, involving undifferentiated individuals in a well-bounded system, permit only a limited range of options, the most obvious of which is privatization of the resource imposed by an agency outside the system (see section B2 below). The second and third formulations, however, as indicated earlier in this chapter, multiply the possible responses and the kinds of agents who make them. The paths that different kinds of agents can pursue depend on their unequal positions within the prevailing economic, social, and political dynamics. Researchers who recognize this situatedness might extend their investigation and refine their accounts until they clarify the appropriate actions or policies to take from the particular positions of the different agents. In this way, awareness of potential knowledge-users can enter at an early stage of the knowledge-making process—well before communicating accounts of the research to others. In short, if we pay attention to the actions and agents that follow from various accounts, we add complexity to the traditional view of the researcher's dialogue with the situation studied. Further work is needed to expose ways in which the audience is implicated in an expanded dialogue that pays attention to the actions or policies of knowledge-users.

B. Socially Situated Researchers Interact with Other Social Agents to Establish What Counts as Knowledge

What can agents do—What social changes can researchers facilitate with their various understandings? (continued)

B1. Simple Formulations: Nonsocial or Social Determination of Knowledge

Researchers are unavoidably engaged with other social agents. When they produce accounts from their research, they are not simply recording their own knowledge, but are trying to persuade others to use those accounts—

to publish, read, accept, build on, apply, and extend them. Traditionally, philosophers of science and scientists themselves maintain that researchers who wish to be persuasive must marshal evidence and order it with concepts, then use their rationally ordered evidence to challenge contrary accounts put forward by others. They contend that social influences have small or transient effects on what is persuasive, and that these effects become negligible by the time a scientific community reaches a strong consensus about a theory, especially when a theory has been used to make further experimental interventions (Hacking 1983). From this perspective, the progression of levels 1 to 4 identified in the last section, through which the system of equivalent units is broken open to expose more complex dynamics, illustrates a standard strategy of using models; namely, start simple and improve by incorporating more factors. Accordingly, the basic schema of Hardin should be seen as an ideal model—reality will differ from it in detail or in more significant ways. One can learn about such reality, however, only by starting with a simple model, comparing it with observations, adding postulates, and progressively improving it. Or, as activists might see it, the simple model draws attention to a problem—degradation of resources held in common—and stimulates the people's involvement, through which they learn more about the complicating details and build experience in implementing policy.

A contrasting formulation of interactions among researchers in establishing knowledge follows from noticing that for many environmental, economic, and political researchers, Hardin's model has not been superseded by accounts of the locally managed commons or of differentiated agents situated in intersecting processes. The coexistence of different accounts of the same phenomena shows that decisive aspects of knowledge-making must lie outside the dialogue with reality, in the domain of social influences. Such influences are especially evident in the interactions researchers have when knowledge becomes controversial (Collins 1981a, 1981b; Nelkin 1984; Martin and Richards 1994).

A commonality between the contrasting formulations is that both amount to system-like interpretations of the practice of researchers. A boundary is posited between the "research proper"—the scientific dialogue with reality—and the social aspects of knowledge production. One domain or the other is the focus of interpretation; its dynamics are coherent and not confounded by the dynamics of the backgrounded domain. In the traditional view, accounts that better *reflect* or *correspond to* reality allow their proponents, through various scientific practices and interactions, to prevail. Such accounts are thus established as knowledge. In the alternative view, social influences inform researchers' practices and interactions and are reflected in the resulting knowledge(s). (Recall the possible narrative in chapter 4, section B, about how Picardi's modeling reflected the technocratic managerial setting in which he worked at MIT.) In both cases, interpreters of knowledge production are located outside the systems they study. They can propose general principles about nonsocial or social determination of knowledge that reduce the complexity of researchers' practices in different fields.

The separation of social and nonsocial aspects of knowledge production is called into question by the open sites described in chapter 2 and the social-personal-scientific correlations described in chapter 3, which led me to investigate the particular complexity of the construction of science from heterogeneous webs of resources in chapters 4 and 5. In the section to follow, I interpret commons research by drawing social-personal-scientific correlations.

B2. Rhetorical Interpretations of Tragic and Locally Managed Commons

In the formulations of section B1, knowledge reflects or corresponds to the reality of nature, to social influences, or—if only transiently—to a mixture of the two. Researchers are treated more or less as vehicles through which reality or social influences determine what knowledge gets established. What, however, do researchers actually *do* in affording persuasive power to these correspondences? One way to address this question is to pay attention to rhetoric, which influences an audience as much by the framing of the case as by its substance—that is, the evidence, logic, or conclusions. The framings of the tragic and the locally managed commons have a variety of effects, which contribute to the social interactions through which commons researchers—now seen as active agents, not as vehicles—attempt to establish what counts as knowledge.

1. Simpling and reinforcing foundational assumptions. Under angle A, I proposed that any Hardin-like state of affairs warrants explanation as a special case of particular intersecting processes (formulation type 3). This is not, of course, the traditional basis for commons discourse, which begins instead with sweeping claims for the general applicability of the simple model of Hardin. Postulates have then been successively added to address the discrepancy between the model and observations, such as distinguishing between open-access commons and those with restricted access or membership. This process of refinement can be interpreted as "simpling":

> Like sampling, "simpling" is a technique for reducing the complexity of reality
> to manageable size. Unlike sampling, "simpling" does not keep in view the
> relation between its own scope and the scope of the reality with which it
> deals. . . . *It then secures a sense of progress by progressively readmitting*
> *what it has first denied.* "Simpling" ... is unfortunately easily confused
> with genuine simplification by valid generalization. (Hymes 1974, 18; my
> emphasis)

In the tragic commons, as in most of economics, self-interestedness is seen as a fundamental characteristic of humans, and this characteristic determines the dynamics of the system. The tragic commons then becomes a result of the "immutable logic of self-interest" (Picardi and Seifert 1976). The belief that the simple model's ideal-type dynamics are fundamental or foundational tends to be reinforced when numerous different variants for different situations are generated. Moreover, the idea that self-interest is human nature is reflected in the very name "tragedy." Classically, a tragedy is something bad that happens to mortals despite their best intentions; only the intervention of the gods can prevent it. Ironically, reinforcement of these foundational assumptions is also effected by theorists of the locally managed commons when they argue that use of non-privatized resources can be governed satisfactorily, provided appropriate social sanctions or regulations are in place *to counteract individual selfishness* (Feeny et al. 1990).

2. Privileging of worldviews and political positions or, more broadly, facilitation of certain social actions or interventions. The assumption of equal, undifferentiated individuals is central to the tragic commons. Given this assumption, the model's dynamics permit only a limited range of options. Hardin explicitly advocates two: privatization of the resource and "mutually agreed coercion." Mutually agreed coercion raises the specter of communism and fascism—recall that Hardin first wrote about the commons in the 1960s—and has not been widely invoked in discussions of the commons. The other options that the model allows are also readily discounted: individuals can leave the system, but this cannot be a solution for every case; individuals can all abandon their desire to accumulate in favor of conservation, but this strategy is undermined by even one holdout or cheat, and so it is never presented as very likely; individuals can drive the system to the inevitable degradation awaiting all non-privatized resources, but this is clearly undesirable. For most Western audiences, privatization remains the only viable option. This message stands, even when the actual record of development efforts casts doubt on the effectiveness of that policy (see note 26).

The tragic commons also conveys less obvious messages. In the actual world, privatization often cements the current claims of unequal individuals—an observation Hardin overlooks. To speak of common resources in terms of the tragic commons, which posits equal, undifferentiated individuals, is to distract attention from the special interests of those with greater claims (Peters 1987). Hardin's model thus makes it easier for powerful interests to get their way. This result was evident in the concessions proposed to secure ratification of the Law of the Sea by the United States, after many years of U.S. opposition during which the tragedy of the commons was regularly invoked. The concessions ensured that the existing seabed claims of U.S. corporations could not be reallocated to the world community (Broad 1994). Once property rights were accepted, discussions of the tragedy quieted (for subsequent developments, see Brown 2001). Over and above unequal property claims, negotiations and contestations among groups with different interests, wealth, and power—the messy stuff of most politics—are omitted from Hardin's picture. The tragic commons thus naturalizes the neoliberal economics of structural adjustment and obscures the politics through which structural adjustment is imposed and implemented in poor, indebted countries.

Discounting politics among unequal individuals is characteristic of two broad orientations toward social action. The enlightened or moral guide instructs listeners how we—an undifferentiated "we"—must change to avoid the impending crisis in question. The technocrat, through an analysis of the scientifically justified or most efficient measures, shows us that it would be in our best interests to submit to those measures. Moralistic or technocratic views of social action are particularly comfortable for those who imagine themselves as the guides and educators, or as the planners and policy advisers. In these roles, they do not have to jeopardize their privileged position through a long-term and necessarily partisan engagement in some particular situation. If they are natural scientists, they can employ their status and skills without retooling for political-economic analysis. The moralistic and technocratic orientations are quite common in environmental discourse (recall Odum's picture of ecosystems as energy circuits in chapter 3; see also Taylor 1997c, 2001d; Taylor and García-Barrios 1997).

Rhetorical privileging of certain political positions can also be seen in the locally managed commons, with its emphasis on designing loosely embedded institutions for managing common resources. For example, officials of the United States Agency for International Development (USAID) invoked the superiority of local institutions to insist that the government in Cameroon adopt free market policies preferred by USAID and drop government-level subsidies and regulation (cited in Moke 1994). Whether or not one endorses the USAID policies, the irony must be noted: the position of the USAID officials—outsiders who overruled the knowledge and institutions of Cameroon officials—was bolstered by their push for local management.

3. Substitution of the exotic for the near-at-hand. Over the last thirty years, environmental degradation and preservation have become familiar issues in affluent countries. In those countries, atomized consumers have found it difficult to organize institutions to ensure that private, corporate, or military property holders bear the full environmental costs of their activities. Concerned consumer-citizens thus have reason to be anxious about their capacity to unite and organize around the goal of influencing corporate and military decision making. In this light, concern about irrationality of non-privatized resources in poor countries can be interpreted ironically, as a displacement from unspoken issues closer to home (Haraway 1989)—issues other than what the tragic commons is *literally* about.

A similar substitution of the exotic for concerns close to home is evident in the attention now given to the locally managed commons in poor countries. Over the last twenty years, dominant political-economic discourses in affluent countries have promoted deregulation, privatization, and decentralization. As the authority of the central state is diminished and economies become more vulnerable to the dictates of globalization or transnationalization, counter-discourses have arisen concerning community and civil society (Agrawal and Gibson 1999; Burbidge 1997). These counter-discourses represent a search for a level of influence intermediate between atomized individuals and the all-powerful market. In this context, the attention given over the last fifteen years to the locally managed commons in poor countries can be interpreted as a displacement to somewhere else of concerns in affluent countries about whether people can still influence social and environmental sustainability.

4. Rendering the special typical. Researchers who argue, contra Hardin, that resources held in common can be well managed often invoke special situations in which the resource and its users are somewhat autonomous from the influence of the government, markets, or industries. This condition is summarized in Ostrom's (1993) design principles of "external acceptance" and "clearly defined boundaries." As studies of such situations accumulate, however, they begin not to appear special, but to be employed to support more general claims. Inquiry that seeks, instead, to define how these situations arise as special cases of more general processes—perhaps non-system-like or intersecting processes—goes to the back burner. When the exemplar becomes defined by such design principles, several things follow.

"Design" connotes that a new institution has to be set up from scratch. This constitutionalist impulse discounts previous history, as if that history becomes irrelevant once a group decides to establish new institutions. As a corollary, little attention is given to the transition problem; that is, how people might engage with ongoing processes to produce the desired social and institutional change. This tendency to discount historical process is evident when discussions highlight success stories and down-play failures, neglecting to analyze the processes leading to both successes and failures. The constitutionalist impulse is also evident when decisions are presented as the result of people coming together to deliberate. This idea of pure politics detracts from the analysis of the political-economic conditions that shape social interactions and their outcomes.

According to the design principles, institutions work better if all the people know and understand each other well. If this principle is used to privilege homogeneous groups, it risks contributing to ethnicization of communities. Such design principles discount the ever-present heterogeneity and inequality within communities. More generally, the idea of an external-internal divide—even in the form of inquiring into how different institutions have been undermined by "external" forces—discounts the ways in which unequal "internal" agents can mobilize and be mobilized by things happening and people operating in far distant places. As observed in section A1 of this chapter, a deep boundary between what goes on inside and who influences the situation from the outside has been a dominant part of research and policy on the tragic and locally managed commons.

Reflection and Transition to Formulation B3
According to the preceding interpretations, the framings of the tragic and locally managed commons have rhetorical effects. My own framing of these

interpretations directs attention to the actions and agents that follow from accounts of the commons—not only literally, but also implicitly. Researchers as rhetoricians influence their audience in some part because of the forms of politics and social action that their scientific accounts privilege.

Of course, my interpretations are overgeneralizations about commons discourse. They provide partial insights—neither necessary nor sufficient—into the practice of commons researchers. Further work is needed to demonstrate the effects of a particular researcher's account on specific audiences and to trace ways in which particular researchers shape their accounts to influence their potential audience. What exactly do researchers do, for example, that allows the models of human nature and social order favored by the audience to get into their accounts? The answer need not refer only to the stage of research at which accounts are communicated to others, but could involve the practice of research from its very outset. In short, further work is needed to specify what knowledge-makers actually do in relation to actual and potential knowledge-users so that commitments to certain social actions get built in their accounts.

B3. Heterogeneous Construction of a Tragic Commons Model

When studying situations in which people attempt to manage resources held in common, researchers take into account the various social situations in which their research is facilitated. If we want an interpretation of this social situatedness that fits into formulation type 3—that is, one that involves dynamics among particular, unequal agents whose actions implicate or span a range of social domains—we need something more specific than the broad correlations evident in the rhetorical interpretations. Instead, we can interpret social situatedness in terms of the diverse practical choices—commitments to certain social actions over others—that particular researchers have to make when making knowledge.

I pursued this interpretive goal in my reconstruction of the heterogeneous webs supporting Picardi's system dynamics computer modeling of nomadic pastoralists—a case of research into the commons (chapter 4, section B). I will recapitulate this interpretation only briefly here: I identified what it would mean in practice to adopt alternatives to each of eight consistent aspects of system dynamics. (The character of these alternatives also matched type 3 formulations of research, i.e., angle A.) This counterfactual analysis gave me insight into the range of diverse resources that

Picardi employed: the available computer software; published data; the short length of time both in the field and for the project as a whole; work relations within the MIT team; the relationship of the United States and USAID to other international involvement in the region; the terms of reference set by USAID and the agency's contradictory expectations of the project; and so on. The practical considerations that were resources for Picardi's knowledge-making were also commitments to certain actions, spanning local and wider social domains. In the diverse, particular ways described, Picardi was affecting social change at the same time he was making knowledge.

Reflection and Transition to Angle C
In examining the interactions of researchers with other social agents to establish what counts as knowledge—angle B—I have interpreted the social situatedness of the researchers in terms of the actions and agents that are built into the knowledge-making process. This emphasis reinforces the suggestion earlier in this chapter that commons researchers consider the relation between their own reasoned understandings and what effects on society result. Yet, from the perspective of heterogeneous constructionism, this reflection is always already happening: all researchers choose particular ways to address the complexities of the situations they study and of their own social situatedness so as to make knowledge and affect social change (see also chapter 4). Therefore, the suggestion from earlier in the chapter needs to be modified: researchers should be more *self-conscious* in the ways in which they jointly address the complexities of situations and situatedness. This challenge adds complexity not only to the practice of research, but also to the interpretation of such practice. Further work is needed if researchers and interpreters are to help researchers take their situatedness less for granted. This is the focus of angle C.

C. Researchers Pursue Social Change by Addressing Self-Consciously the Complexities of the Situations They Study and Their Own Social Situatedness

What can agents do—What social changes can researchers (including interpreters of science) facilitate with various understandings of complexity? (continued)

As I am construing the term, "social change" can include something as modest and local as stimulating change in the concepts used by members of one's audience (this chapter, section A, and the prologue). For example, my account of the heterogeneous construction of Picardi's modeling of nomadic pastoralists (this chapter, section B3; chapter 4, section B) was intended to influence interpreters of science to think more about the implications of the observation that agents involved in building models of socio-environmental complexity draw on diverse components from a range of domains of social action. This account could also be read as an invitation to socio-environmental scientists, including commons researchers, to interpret their own research or to pay attention to interpretations made by others.

A slightly less modest form of social change could happen if some scientists used such interpretations to modify the social and technical conditions involved in their knowledge-making with a view to producing a different account of the complexities of the situations they study. For example, my account of Picardi's modeling exposed multiple points of engagement at which the modeling project could have been modified—specific points at which concrete alternative resources could have been mobilized in order to analyze complex situations without modeling them as well-bounded systems. Socio-environmental researchers in situations comparable to the MIT study might be able to draw lessons by analogy about ways to alter their personal, scientific, and social facilitations and so modify the directions of their research.

"Lessons by analogy" is an apt description because significant translations would be necessary to relate my interpretation to a new situation. I would not expect the same set of practical considerations or resources to transfer to the interpretation of other cases of research on the tragic commons, let alone to that of cases of research into locally managed commons. I would not even expect the same analysis to emerge if different interpreters were to consider Picardi's modeling from another entry point; say, around the concerns of USAID, which was trying to use the MIT study to elevate the role of the United States in West Africa. In seeking to expose the diverse practical considerations with which Picardi dealt, other interpreters would face their own set of practical considerations, which would differ according to the particular social situations in which each worked.

In practice, the additional complexity of the situatedness of researchers is usually discounted or backgrounded as they pursue other forms of social change, such as building careers and institutions, using transforming language, facilitating policy formulation, or enlisting support in some environmental or scientific controversy. Nevertheless, I do not agree that reflection on one's situatedness is irrelevant to changing it (Fish 1989) because, from the perspective of the heterogeneous construction of science, reflection in relation to decisions about practice is always already happening (this chapter, section B3; chapter 4). In what ways, then, could researchers be more self-conscious about understanding *and* changing the complexities of their situatedness *and* of the complex situations they study?

C1. Moves to Discount or Background Complexity

In this section, I run through a number of ways in which researchers who self-consciously address their own social situatedness or the complexities of the situations they study might decide to suppress the linkages among angles A, B, and C or to suppress the dynamics among particular, unequal agents whose actions implicate or span a range of social domains (i.e., to produce type 1 or 2 formulations instead of type 3 formulations). The sixth move—introducing opening-up themes—has been pivotal in the framework of this chapter.

1. Shaping accounts to be transmitted and digested. Researchers who think about communicative and cognitive constraints on how much complexity others can absorb may, in the interests of influencing an audience, shape their own accounts so as to be, first, *transmissible*, which usually means preferring type 1 over type 3 formulations and angle A over an integration of angles A, B, and C; and, second, *digestible*, which necessitates departing as little as possible from the formulations of other researchers in one's audience. Conventionally, these formulations center on marshaling concepts and evidence about the situation studied (angle A) and take most aspects of the researchers' social situatedness in relation to their knowledge-making (angle B) for granted. (Similarly, if the interpretation of commons research—perhaps by the researchers themselves—is viewed as an angle A′, the corresponding angle B′—the social situatedness of the interpreters—is left unexamined; see chapter 5, section A.) It follows that, if researchers are expected not to refer to their own situatedness, then they need not examine the audience's situatedness in relation to their capacity to be influenced by the researchers' knowledge-making (angles B and C).

2. Drawing clear political lines. The interest in political impact that some researchers have elicits similar thinking about what makes an account readily transmissible and digestible. Simple themes, such as "common resources need to be privatized to prevent degradation" and "population growth will lead to environmental degradation," appear to provide effective rhetoric in larger social mobilizations. Similarly, interpreters of science mobilize around simple themes such as "scientists can be best understood by reconstructing the context in which they worked" or "democracy requires public participation in debates over scientific claims" (Lynch 2003). Researchers who take note of the currency given to simple themes and who want to enlist others to support their position on some political or social cause may choose to discount or background additional complexity. Typically, their accounts draw distinct lines between positions and identify the one they hold to be factually (or morally) right. Differences within each position are treated as a secondary consideration; such a move reduces the attention paid to differences among people in the ways they come to their positions—that is, to differences among the heterogeneous resources people mobilize in their knowledge-making.

Some politically inclined researchers construe the move to draw clear political lines as more than a matter of rhetorical shaping of text and argument; they support the move in realist terms. They argue that misrepresentations of reality (or unjustifiable moral arguments) underwrite social practices to which they are opposed, and they claim that their politics would be weak without highlighting such connections. In the terms of this chapter, such realists accept a simple formulation of angle B: they discount the dependence of their factual (or moral) claims on the many other resources that researchers and other members of their audience use in negotiating their own contributions to changing knowledge, society, and the environment (Sismondo 1993b; Taylor 1995b).

3. **Presenting accounts of complexity to stimulate others to address their situatedness.** Even when researchers' accounts do not delve into the complexities of their own social situatedness, they may prod members of their audience to address social situatedness for themselves (chapter 5, sections A and C). When I use the intersecting processes account at the end of the classroom simulation that begins this chapter to address people who accept the tragic or locally managed views of the commons, I hope to convince some of them that type 3 formulations best capture the dynamics of the use of common resources. The virtues of their own intersecting processes accounts might then lead these "converts" to mobilize collaborators, sources of funding, and so on in order to shift the orientation of their research, policymaking, or other practices. Similarly, when, as an interpreter of science, I formulate accounts of the politics or social actions favored by or built into researchers' socially situated knowledge-making (i.e., angles B2 and B3, respectively), I hope that the virtues of those interpretations might lead readers to mobilize new resources in order to modify their own work (Taylor 1997c, 1997a, 210ff). In both instances, attention is

drawn to the complexities of the situations studied, but the scientific or interpretive "dialogue" with those situations (angle A) is meant to promote change in the audience's knowledge-making and other practices.

4. Experimenting and taking stock during long-term engagement. Researchers who are prepared to engage with others for extended periods around particular environmental and scientific controversies have the opportunity to experiment, take stock, and adjust. They can do so knowing that they are guided by (explicit or implicit) models that provide simple and partial accounts of the complexity of the situations they are engaged in— complexity that is a composite of the situation the researchers are studying and the situatedness of the various agents involved. Experimentation and ongoing assessment guided by partial models is, more or less, the spirit of Adaptive Environmental Management (see narrative at the end of chapter 4; note 19; and the epilogue).

5. Offering multiple angles of illumination. Even without long-term engagement with others around particular controversies, researchers who think about what is readily transmissible and digestible need not discount the complexity of researchers' social situatedness (angles B and C) or of intersecting processes and unruly complexity (type 3 formulations) as much as in moves 1–3 (above). Suppose any simple (type 1) formulation is reconceived as one of many angles of illumination on a situation, or as a theme to be woven together with other themes, but not as an explanation on its own. The resulting picture can be more complete than implied literally by the factors, relations, and boundaries of the different simple formulations. This is often the spirit of interpretive social sciences and history. The reflexive turn in those fields, loosely associated with postmodernism, adds further angles of illumination on the researcher's own situatedness.

Two problems emerge, however, when simple formulations are employed as angles of illumination: First, how can researchers weave multiple themes into a more complete picture (a problem raised in chapter 2 in the context of ecological modeling)? Second, how can researchers sustain the heuristic quality of themes; that is, not allow bounded and system-like dynamics to anchor one's account of complex situations (Taylor 1997a, 210ff) or overlook the special conditions needed for such dynamics to apply (recalling chapter 1, section B)? These problems are to some extent addressed by the next move, which is the pivotal element in the framework of this chapter.

6. **Introducing opening-up themes.** Researchers can use themes that build in a persistent opening up of issues—that point to the hidden complexity of simple formulations and to further work needed to address the differentiated detail and other complexities of particular cases. Such themes are my type 2 formulations. Because they have an orientation toward further work (to produce type 3 formulations), it is harder to slip back into simple, system-like (type 1) formulations or to focus on angle A as if research "proper" were separate from angles B and C. At the same time, such themes are intended to be more transmissible and digestible than particular accounts of intersecting processes. The tensions between type 1 and type 3 formulations are kept in view and can be productive of further work. Table 6.1 summarizes the opening-up themes introduced in this chapter; see also the summary of themes and questions opened up at the end of the book.

7. **Drawing the audience into experiential activities.** People take in more of the material presented to them when they practice using it and connecting it to an area of their own interest. Following this principle, researchers can involve members of their audience in experiential activities. For example, at times I have presented abbreviated versions of this chapter's three types of formulations, then led audiences through a set of exercises in which they identified an example of a simple formulation of a topic from their own teaching or research, invented an opening-up theme, and sketched a corresponding particular and differentiated account (Taylor 1998a). The experience of such activities is, by definition, difficult to convey in a text (Taylor 2002b). Equivalent experiential sessions can be envisaged that would concern researchers' situatedness (angle B).

Reflection and Transition to Formulation C2
These moves that discount or background complexity involve few details about particular, differentiated agents. The emphasis is on individual researchers who produce accounts so as to have an effect on their audiences. In that sense, C1 formulations are system-like—the researcher is the inside of the system, and the audience is the outside "environment" whose characteristics are more or less given. Agency appears "concentrated" inside individual researchers.

At the same time, the moves can be said to affect social change only if they affect the audience. In other words, the moves all rely on some response from members of the audience, who are socially situated in their different and particular ways. Agency is implicitly "distributed" to them;

Table 6.1 Opening-up Themes

- A qualitative change in the analysis of causes and in the implications of the analysis can result if
  1. the system's dynamics—the rules and system structure—are modeled as mutable;
  2. social situatedness is considered primary;
  3. boundaries are treated as permeable; and
  4. the paths individuals can pursue are analyzed in terms of unequal individuals subject to further differentiation as a result of their linked economic, social, and political dynamics.

($A2 \rightarrow A3$)

- A qualitative change in the analysis of causes and in the implications of the analysis can result if attention is given to the special conditions necessary for a simple model to apply to a situation. ($A2 \rightarrow A3$)
- A qualitative change in the analysis of causes and in the implications of the analysis can result if
  1. structures are seen as subject to restructuring;
  2. boundaries and categories are seen as problematic; and
  3. control or generalization is seen as difficult.

($A3 \rightarrow$ further work needed under a type 3 formulation of angle A; see also the larger set of contrasts in chapter 5, section C)

- Making and using knowledge are influenced by researchers' simultaneous dialogues with other knowledge-users and with the situation studied. (Reflection on both versions of $A \rightarrow B$)
- The concern of commons researchers with the ways in which the actions of rational agents contribute to undesired collective, social outcomes can be extended reflexively to ask what social change researchers affect or facilitate with their various understandings. (Introduction to alternative version of A; prefigures C)
- Attention to the actions that follow from different accounts can bring awareness of potential knowledge-users into the knowledge-making process— well before the point at which the resulting accounts of the research are communicated to others. (Reflection on alternative version of $A \rightarrow B$; prefigures C)
- Over and above the effect of what researchers' accounts literally state, the framing of accounts has several rhetorical effects on the audience:

Note: Corresponding sections of Chapter 6 are indicated in parentheses; "A2 $\rightarrow$ A3" denotes "section A2 points to or opens up section A3."

(Continued)

Table 6.1 Opening-up Themes—continued

1. Simpling secures a sense of progress by readmitting what it had first denied, and thus reinforces foundational assumptions.
2. Simple, undifferentiated models privilege certain political orientations or courses of social action.
3. Concerns about exotic situations are displacements of the concerns closer to home that the researchers have.
4. Taking special situations as exemplars distracts attention from the historical processes that produced both the special situations and the others.

(B2 → the linkage of angles A and B, and, in combination with the following themes, → B3)

- Certain courses of action and kinds of agents are facilitated over others not only in the framing of accounts, but also in the ways in which the research is formulated in the first place. (Reflection on B2 → B3)
- Particular researchers take into account at the same time the situations they are studying and the social situation in which their knowledge-making is facilitated, making diverse practical choices—commitments to certain social actions over others—when making knowledge; this process can be undertaken more self-consciously. (B3 → C)
- The moves by which a researcher discounts or backgrounds the interwoven complexities of situations and situatedness (C1) can also be seen as an invitation to members of the audience to mobilize resources and make diverse practical choices so that they can digest the researcher's account and do something with it. (Reflection on C1 → C2)
- Researchers know more than they are prepared to acknowledge until encouraged or prodded by interaction with others. (Reflection on C2 → C3)
- Knowledge-making, social-changing agents move or vibrate among their own variants of chapter 6's three types of formulations and three angles. (Reflection on C2 → C3)
- Workshop conveners need to integrate concentrated and distributed views of agency when they attempt to modify or restructure workshops with the aim of helping reflexive researchers modify or restructure their situatedness. (Reflection on C2 → C3)
- In order that workshops not discount heterogeneity and inequality, a wide range of researchers and social agents need to be brought into interaction and kept working through differences until plans and practices are developed in which all the participants are invested. (Reflection on C2 → C3)

Table 6.1 Opening-up Themes—continued

- To facilitate a culture of participatory restructuring of the distributed conditions of knowledge-making and social change, more work is needed on what agents can do—but not alone or solely through their accounts of the world—to contribute self-consciously to the ongoing restructuring of the dynamics among particular, unequal knowledge-making agents whose actions implicate or span a range of social realms. (C3 → work desired of readers to fully realize C3)

they are left to mobilize resources and make diverse practical choices to digest the researcher's account and do something with it. This dualism of concentrated and distributed agency points to a challenge for researchers: Instead of distributing agency implicitly, can they do so self-consciously (as moves 3, 6, and 7 in section C1 foreshadow)?

C2. Mapping Workshops: A Self-Conscious Move to Distribute Agency

The notion of distributed agency highlights the range of different social agents involved in a situation and the diversity of resources they mobilize. In the locally managed commons, the agency of users is more distributed than in the tragic commons; with differentiated agents situated in intersecting processes, it is even more so. Distributed agency also applies to researchers who employ heterogeneous resources drawn from a range of intersecting social domains. Picardi's modeling work, for example, could have been modified, but this would not have followed readily from a mere change of worldview on his part; instead, many practical considerations and social negotiations would have been involved (chapter 4, section B; chapter 5, section A). Indeed, an interpretation that teases open the heterogeneous construction of research points, as does a socio-environmental account of the intersecting processes (A3), to multiple, particular change-inducing interventions or engagements (chapters 4 and 5).

Mapping workshops, described in chapter 5, section B, take the notion of distributed agency one step further than intersecting processes accounts and heterogeneous constructionist interpretations. In mapping, it is not some second party who does the interpretation. Researchers themselves, stimulated by interactions with other workshop participants, attempt to interpret their own heterogeneous webs. They reflect explicitly on their own situatedness in relation to their ability to study the situations that interest them, and they identify multiple potential sites of engagement and change as knowledge-makers (see also Taylor 1990, 1999b).

Reflection and Transition to Angle C3
A built-in limitation of mapping workshops points to ways to take the notion of distributed agency even further. Workshop interactions center around maps or other representations made by *individual* researchers and thus preserve a degree of concentrated agency. Indeed, this tendency is hard to avoid, given that workshops are meant to encourage researchers to address *self-consciously* the situations they study and their own social situatedness so as to affect social change. Building on this observation, let me reflect on the experience of mapping workshops in relation to this chapter's questions about linking understanding and self-conscious agency (see also chapter 5, section B).

1. **Vibrating agency.** In map-making and mapping workshops, participants articulate connections that had previously been unexamined, unspoken, or discounted. That is, when encouraged or prodded by interaction with others, researchers show that they know more about their situatedness than they had been prepared or able to acknowledge. It follows, however, that day-to-day practice and discourse must be discounting researchers' awareness of their situatedness (Taylor 1993, 1997a). This means, according to the perspective of heterogeneous constructionism, that simple, system-like formulations about one's social situatedness are serving as resources for agents in their knowledge-making. For example, the motivation or drive Picardi expressed for demonstrating a case of the tragedy of the commons (chapter 4, section B; chapter 5, section A) was undoubtedly one of the diverse resources he mobilized.

The tension between, on one hand, what is acknowledged and stated and, on the other, what is known and could be acknowledged suggests a reconciliation of concentrated and distributed agency. Knowledge-making agents must always be moving or "vibrating" among their own variants of this chapter's three types of formulations and among the three angles for viewing their own practice. If such vibrating characterizes, albeit schematically, what agents always do, then mapping workshops are not making researchers do something new. Instead, they are merely strengthening the vibrations in the direction of self-conscious knowledge-making and social changing (angle C) and in the direction of attending to particular dynamics (type 3 formulations). Further work is needed to acknowledge such vibrating agency and mobilize it in productive ways.

2. Modifying or restructuring conditions. Researchers' decisions to participate in mapping workshops point to a concentrated notion of agency, but the conditions needed to enable researchers to modify or restructure their situatedness are obviously more distributed. Workshops could never provide all of those conditions, but workshop processes can stimulate and support the concentrated agency that goes into reflexive scientific practice. Through experiment and experience, workshop conveners and facilitators might arrive at workshop processes that lead to more striking changes among participating researchers than achieved in the initial two mapping workshops. Yet, as noted at the end of my earlier discussion of those workshops (chapter 5, section B), whether the processes and interactions can be established and sustained until their full ideal is achieved is something that depends on more than a workshop convener's will. The conditions for powerful workshops are more distributed. Further work is needed to acknowledge in productive ways this tension between concentrated and distributed views of agency in modifying or restructuring conditions for modifying or restructuring researchers' situatedness.

3. Changing collaborations. Distributed agents depend on a diversity of resources, many of which depend on the actions of others with whom their work intersects. Recall, however, that the participants of the pilot workshops were self-selected by their willingness to commit time to reflect on their research and possible future directions. Mapping workshops do not ensure heterogeneity and inequality if they attract participants who have decided to engage with one another in more or less intentional groups. (Ironically, this observation recalls my critique of "design principles" for locally managed commons; see section B2 of this chapter.) This shortcoming points to the challenge of bringing into interaction not only a wider range of researchers, but a wider range of social agents, and to the challenge of keeping them working through differences and tensions until plans and practices are developed in which all the participants are invested.

4. Translocality. A wider range of participants would undoubtedly ensure that the maps (or other accounts of situations and situatedness) were more diverse. However, the experience of mapping workshops suggests that in each of these accounts, the view of situatedness might still be individual-centered. The challenge, if the actions of "particular, unequal agents" are to be seen to "implicate or span a range of social domains," is to incorporate analysis of changes that arise beyond the local situation or appear at larger temporal and spatial scales. Workshop leaders could draw attention to empirical regularities, such as increased public support for environmental research or constraints on national environmental regulation imposed in the name of free trade. Workshop leaders could also discuss correlations between knowledge and social interests, more abstract theories about the dynamics of social change, or the role of agents in producing and reproducing social structuredness,[27] including its gendered and racialized dimensions.[28] At the same time, they do not want to suppress researchers' reflection on the particular situations in which they do their research. Further work is needed to develop ways to bring the *translocal* into workshop interactions while still preserving the local and the personal.

C3. Facilitating Processes of Participatory Restructuring

Taken together, the preceding reflections on the experience of mapping workshops suggest what the final component of my 3×3 framework should be; namely, a case in which the initiative of the workshop or project convener meshes with the circumstances and achieves the ideal of bringing together a diverse set of researchers and other social agents, allowing them to acknowledge and mobilize their vibrating agency, injecting themes about social structuredness, and sustaining everyone's interaction until new complexity-addressing collaborations emerge. However, if I could point to such a case and used it to influence readers, this would amount to the move 3 described in section C1—I would be providing an account of the complexities of the situation studied, but leaving it to readers to mobilize the collaborators, sources of funding, and so on needed to contribute to work that matches the ideal. For readers to expect an author to provide such exemplary cases, and for the author to expect readers to take those cases as models to follow, would be to place a lot of weight on the concentrated agency of author and readers alike. Instead, taking note of the vibrations between distributed and concentrated agency, I propose a C3 that leaves *open* and *active* questions about the roles of individuals and their knowledge, themes, and other awareness of complex situations and situatedness.

The final component of the 3×3 framework is thus a question mark (see fig. 6.3). The unfilled cell is intended as an opening-up theme, reminding readers of the challenge to which the eight components A1–C2, taken together, point; namely, the challenge of participating with others in restructuring the distributed conditions of knowledge-making and social change.

With a question mark as the final component, the chapter as a whole suggests two "inversion" themes that point to additional complexity and further work needed. Conventionally, scientific *process* is presented as a transient state on the path to the *product*—social agents engage with other knowledge-using and knowledge-making agents in the course of establishing knowledge and representations. If this perspective were inverted, interpreters of science could view an emphasis on product as a contribution to a particular process; namely, one that discounts or backgrounds the ways in which researchers affect social change by addressing the complexities of situations and situatedness. Moreover, researchers and interpreters tend to judge a product against the ideal of a complete theory that, once it has been established, dictates the correct paths to take in knowledge-making and social change. Inverting this perspective, researchers and interpreters could see themselves as assembling and drawing from a growing toolbox of themes in the never-ending process of negotiating the contingencies of work and life. Then it would be only at a distance that contingent use of themes could be made to appear systematic or system-like.

This chapter's 3 × 3 framework and the idea of formulating opening-up themes are two recent additions to the box of tools that I use in trying to nudge audiences in the direction of attending to particular, distributed dynamics and of contributing self-consciously to knowledge-making and social change. Equally important for that nudging has been my learning new tools for facilitating group interaction and reflexive practice, then sharing those tools by applying them in workshop settings and teaching.[29] However, whether I am emphasizing concepts or experience-in-practice, I try to keep in view the particularity of my limited contribution. Clearly, more work is needed on what I and other agents can do—but not alone, nor solely through our accounts of the world—to contribute self-consciously to the ongoing restructuring of the dynamics among particular, unequal knowledge-making agents whose actions implicate or span a range of social domains.

Epilogue: Three Stories

The conceptual exploration presented in this book began with the question of how ecologists could account for order arising out of the complexity of situations that build up over time from heterogeneous components and are embedded within wider dynamics, and in which there is ongoing restructuring—what I have come to call unruly complexity. An important aspect of the progress I have made toward answering this question is a shift in emphasis from the word "account" to the word "how"—from representations of complexity to representing-engaging—from product to process. At the beginning of the journey, I envisaged that an answer would take the form of a theory or models that provided an explanation of ecological complexity. At the end, I am inviting researchers who want to reconstruct the unruly complexity of ecological and social situations to become more self-conscious about their engagement within the complexity of the situation studied and of the social situations that enable them to do their research. The intersecting ecological, scientific, and social processes in the work of researchers involve diverse components and agents and span a range of spatial and temporal scales—the boundaries of unruly complexity are problematic. As both a conceptual and a practical matter, the framework of the last chapter had to leave as an "exercise for readers" the challenge of using your knowledge, themes, and other awareness of complex situations and situatedness to contribute to "a culture of participatory restructuring of the distributed conditions of knowledge-making and social change."

With this ending, the book as a whole becomes an opening-up theme. The book does not provide a theory to explain unruly complexity in any specific field or situation, but opens up issues about addressing complexity in ways that point to further work that needs to be undertaken to deal with particular cases. On some occasions, I have attempted to motivate this theme in the space of a single lecture through a rapid presentation of the framework of the last chapter. On other occasions, however, I have found myself adopting an approach that amplifies the moves in section C1 of chapter 6; namely, to use certain stories to convey some meaningful things that researchers might work on with and within the framework. Although I have been wary of ways in which the narrative form tends to reinforce our experience of ourselves as concentrated agents,[30] I am learning that stories like those that make up this epilogue can keep distributed agency in view as we seek to grapple conceptually and practically with unruly complexity ...

◆ ◆ ◆

A. Participation

I now have an image of critical reflection on practice in which ecological researchers and interpreters of science are able to respond to developments—predicted and surprising alike—by continually reassessing their knowledge, plans, and actions, as well as the engagements that make them possible. As mentioned earlier, this is an ideal inspired by participatory action researchers, who shape their inquiries through ongoing work with and empowerment of the people most affected by some social issue.[31] Since I learned about a particular participatory Kenyan agroforestry project in 1988, I have often spoken of it to others, using the case to illustrate the ideal of bridging the divide between an outside analyst and the subjects whose social and ecological situation is being analyzed.

In the mid-1980s, CARE, an international aid and development organization, decided to respond to the excessive removal of trees in agricultural areas in western Kenya. It embarked on a project to establish an extension system that would promote and provide support for tree planting by farmers on their holdings. CARE sought to overcome the shortcomings of previous agroforestry projects in the Sahelian region of Africa, which had largely failed—one estimate of the average cost those projects had incurred for each surviving tree was $500. At the same time, CARE wanted a research component built in to analyze systems of farm production—not only of crops, but also of things necessary to basic household needs, such as energy, shelter, and water. The research aimed to tease out the trade-offs, constraints, and benefits of growing trees within those systems (Vonk 1987).

The leaders of this development project, agroforesters Remko Vonk and Louise Buck, identified one reason for the previous failures: the community-based nurseries and plantations of previous projects had left the beneficiaries of the tree products and timber ill-defined (Vonk 1987).

Many of the local participants saw the tree planting as someone else's project, and thought the benefits would be unlikely to come their way. Vonk and Buck reasoned that if trees were planted on individual farms, the ownership would be clearer; the local Kenyans implementing the project would also be the ones reaping the benefits. Moreover, the project leaders aimed to facilitate local participation in the design and evaluation stages of the project. In pursuing this goal, they drew on their experience in a pilot project and on the experience of others in previous health care extension projects.

This combination of local and outside influence characterized the project as it developed. First, CARE entered only those farming communities that invited it. Initial interviews were conducted to learn about the existing uses of trees on and off the farms: Which trees are being used; which had been used; which could be used? What are the reasons for not planting trees? Much of the interviewing was conducted by extension workers, whom CARE directors trained not to transmit information, but instead to "Respect, Encourage, Ask, and Listen." In response to information emerging from the interviews, CARE's preliminary plan of planting four species was modified to allow for selections from a menu of forty-eight species. The techniques of cultivation that the researchers adopted, using indigenous systems as a starting point, were understandable to the farmers and could be managed by them within their labor and other seasonal constraints. In turn, the extension agents' connection with farmers helped them plan, monitor, collect data on, and analyze the different tree-planting arrangements.

The resulting agroforestry practices and results differed markedly from those of previous systems and from the approaches of CARE's agroforestry specialists, which had focused on trees that would directly serve agriculture, for example, by fixing nitrogen and making it available in the soil. The case of *Markhamia platycalyx* is illustrative. This species, virtually unmentioned in the agroforestry literature, was the most commonly found species in cropland in the district. The tree did not enhance crop growth, but, as interviews with the farmers revealed, *M. platycalyx* grew quickly and so was used to demarcate family compounds and plots. Reduction in crop production because of shading and root competition could be minimized if the trees were pruned regularly. The leaves became a source of mulch and compost, and scattered trees contributed to soil conservation and had a windbreak effect that protected the crops in the fields. The trees could be cut for poles when cash income was needed. They could also be used to provide timber or shade. Finally, the leaves were used in preparing food and in medicines. CARE research confirmed that farmers generally knew how to manage the species well for these different uses. At the same time, CARE was able to help the farmers by contributing research results on the optimal time for harvesting of trees to be used for poles and on possible causes of seedling death.

In general, the trees that farmers favored turned out to have the following characteristics: They tended to require little management. They were intercropped with crops or even interspersed throughout the fields; they were not only planted as hedgerows. Their products, such as firewood and poles for building, sometimes compensated for the negative effect they had on the yields of adjacent crops. Over and above these characteristics, other factors influencing the use of different tree species on particular farms or more generally included the histories of different farms—in particular, where family compounds had been abandoned, leaving their traces in nutrients from feces and ashes, and how land had been subdivided among sons; the different needs of men and women; and the need for firewood in areas close to Lake Victoria in order to smoke or fry Nile perch (a species that, unlike the fish it has displaced since being introduced to the lake in the 1950s, is too oily to be sun-dried).

CARE's project involved researchers' collaboration not only with farmers, but also with community groups. For example, researchers worked with schools to establish seedling nurseries. When removal of seeds by termites became a problem, the project leaders insisted that pesticides not be used near schoolchildren and sought nontoxic solutions. Some control schemes suggested by the community members failed, but success was eventually achieved following some farmers' recommendation that seeds be surrounded with ashes. Again, in the spirit of collaboration, one CARE official's innovation of using plastic to avoid dampening the ashes when watering the crops reduced the number of times the ashes had to be reapplied.

This combination of local and outside influences occurred in many other varying ways. The extension workers CARE trained were young adults from the area, who would continue to live and work in the area after CARE withdrew. Yet CARE deliberately chose to train women and men in equal numbers, which would not have occurred if selection had been subject to the unequal gender norms of the community. CARE allowed local practices to form the focus of their research, but the CARE agroforesters also made observations and conducted trials to relate the seedling survival rates, growth rates, nutrient contributions, and cash values of products of different species to soils, planting densities, pruning and harvesting practices, and so on. The results of these investigations informed the advice they gave to the local farmers and to agroforesters in other areas of the Sahelian region.

CARE's emphasis on achieving meaningful local participation stemmed from an awareness that a successful project would require a complex set of negotiations involving the project's funders and government bodies. Indeed, CARE deliberately located this project in an area without significant involvement by government forestry workers so that it could become established and visibly successful before it incited bureaucratic interference. In retrospect, CARE officials concluded that if this project were to be taken as a model for other areas, and if the extension networks they had established were to remain viable, they needed more government endorsement than they had sought. This reservation aside, the participatory approaches of subsequent CARE projects in agriculture, forestry, health care, and other areas drew heavily on the model of the Kenyan agroforestry project. The success of the agroforestry project was evident when, during the evaluation process, the farmers were asked: "Who decided which species to grow? Who owns the production process?" The answer to both questions was clear; the farmers exclaimed: "These trees are ours!"

B. Flexible Engagement

In the late 1990s, I attended some facilitation training at the Canadian Institute of Cultural Affairs (ICA). ICA's techniques have been developed through several decades of "facilitating a culture of participation" in community and institutional development. Their work anticipated and now exemplifies the post-Cold War emphasis on a vigorous civil society; that is, on institutions between the individual and, on one hand, the state and, on the other hand, the large corporation (Burbidge 1997). ICA planning workshops involve a neutral facilitator leading participants through four phases: practical vision, underlying obstacles, strategic directions, and action plans (Stanfield 2002). These phases mirror and make use of the "objective, reflective, interpretive, decisional" steps of shorter ICA "focused conversations" (Stanfield 1997). The goal of ICA workshops is to elicit participation in a way that brings insights to the surface and ensures that the full range of participants are invested in collaborating to bring the resulting plans or actions to fruition (Taylor 1999a).

Such investment was evident, for example, after a community-wide planning process in the West Nipissing region of Ontario, 300 kilometers north of Toronto. In 1992, when the regional Economic Development Corporation (EDC) enlisted ICA to facilitate the process, industry closings had increased the traditionally high unemployment rate to crisis levels. As well as desiring specific plans, the EDC sought significant involvement of community residents. Twenty meetings with over 400 participants moved through the first three phases—vision, obstacles, and directions. The results were synthesized by a steering committee into common statements of the vision (fig. E.1), challenges, and strategic directions. A day-long workshop attended by 150 community residents was then held to identify specific projects and action plans and to engage various groups in carrying out projects relevant to them.

In a follow-up evaluation five years later, the EDC found that it could not simply check off plans that had been realized. The initial projects had spawned many others; indeed, the EDC had been able to shift from the role of initiating projects to that of supporting them. It made more sense, therefore, to assemble the accomplishments under the headings of the original vision and strategy documents. Over 150 specific developments were cited, which demonstrated a stronger and more diversified economic base and a diminished dependence on provincial and national government social welfare programs. Equally importantly, the community now saw itself as responsible for these initiatives and developments, eclipsing the initial catalytic role of the EDC-ICA planning process. Still, the EDC appreciated the importance of that process and initiated a new round of facilitated community planning in 1999 (West Nipissing Economic Development Corporation 1993, 1999).

When I learned about the West Nipissing case, I could not help contrasting it with my own experience in the Kerang study (chapter 4, section A). In that case, a detailed scientific analysis was conducted at some distance from those directly affected by the problems of salinization and economic decline. Projections of the economic and ecological future were straightforward as long as they preserved the basic structure of the situation. When innovative possibilities, such as reforesting abandoned land, were considered, the analysis became difficult. The audience for the final analyses was small and attention to the report short-lived. The Ministry was unable to implement the policy change it desired, and nothing more then became of the two or three person-years of research.

The West Nipissing plan, in contrast, built from straightforward knowledge that the varied community members had been able to express through the facilitated participatory process. The process had been repeated, which presumably allowed the participants to factor in changes and contingencies, such as the start of the North American Free Trade Agreement and the decline in the exchange rate with the United States.

WEST NIPISSING VISION

Vision 20/20 — February 1993

STRONG DIVERSIFIED ECONOMIC BASE			EXCITING ATTRACTIVE COMMUNITY TO LIVE IN			ACTIVELY INVOLVED POPULATION		
WIDELY PROMOTED TOURISM BASE	EXPANDED BUSINESS DEVELOPMENT	APPROPRIATE NATURAL RESOURCES DEVELOPMENT	WELL MAINTAINED EXPANDING INFRA-STRUCTURE	COMMUNITY BASED SERVICES	RESPONSIVE ACCOUNTABLE UNIFIED GOVERNMENT	ACTIVE INVOLVED COMMUNITY	IMPROVED RECREATION OPPORTUNITY	LIFELONG EDUCATION FACILITIES
Broad Based Tourism Promotion	Modern Recycling Facilities	Forestry Development	Improved Transportation Network Locally/Area	First Response Teams	Effective Cooperation Between Municipalities	Active Involvement of Citizens in All Community Developments	Youth Activities Promoted and Supported	Accessible Expanded Adult Education
	Northern Ontario Service Industry Centre			Community Based Services for Mental Health & Physically Challenged		West Nipissing Team Cooperation		
Improved Four Season Accommodation	Appropriate Natural & Resource Based Industry	Expanded Local Agricultural Market		Expanded Local Access to Specialized Clinics	Ongoing Citizen Involvement in Local Government	West Nipissing Friendly Welcoming Community	Improved Access to Lake Nipissing	Focused Job Training Programs
Accessible Waterways and Waterfronts	Incentive Programs to Attract Businesses	Fish Hatcheries	Well Serviced Community	Coordinated Integrated Services under One Roof	Local Service Boards in Unincorporated Municipalities	Rural Residential Development	Broadened Leisure Activities and Facilities	
Packaged Tourist Attractions & Tours	Francophone Bilingual College		Environmentally Responsive Community	Expanded Vibrant Senior Citizen Community		Open Communication across West Nipissing		Enhanced Post Secondary Education
Expanded Coordinated Community Festivals	Local Businesses meet all needs	Clean Lake Nipissing		Restructured Social Assistance System	Re-evaluate Land Use By-laws	Youth Involved in Planning All Activities	Improved Organized Sports	
	Attract Government Offices							

Figure E.1 The vision for the West Nipissing region in Ontario, Canada, produced by a community-wide planning process in 1992. (From West Nipissing Economic Development Corporation 1993.)

And, most importantly, the process had led community members to become invested in carrying out their plans and to participate beyond the ICA-facilitated planning process in shaping their own future.

Some difficult questions were opened up for me by this contrast, given that my own environmental research has drawn primarily on my skills in quantitative methods. What role remained for researchers in inserting the translocal—that is, their analysis of changes that arise beyond the local region or at a larger scale than the local—into participatory planning? For example, even if I had moved to the Kerang region and participated directly in shaping its future, I would still have known about the Ministry's policymaking efforts, the data and models used in the economic analysis, and so on. Indeed, the "local" for professional knowledge-makers cannot be as place-based or fixed as it would be for most community members. What would it mean, then, to take seriously the creativity and capacity-building that seems to follow from well-facilitated participation, but not to conclude that researchers should "go local" and focus all their efforts on one place?

When I first presented the West Nipissing–Kerang contrast (Taylor 2000b), I asked the audience to explore this question through some guided freewriting (note 29; chapter 6, section C1, move 7). Out of my own free-writing on that occasion, a new term emerged: "flexible engagement." This term seemed to capture the challenge for researchers in any knowledge-making situation of connecting quickly with others who are almost ready to foster—formally or otherwise—participatory processes and, through the experience such processes provide their participants, contributing to enhancing the capacity of others to do likewise. The term plays off the "flexible specialization" that arose during the 1980s, wherein transnational corporations directed production and investment quickly to the most profitable areas and set aside previous commitments to full-time employees and their localities. Would flexible engagement constitute resistance or accommodation to flexible specialization? This remains an open question.

C. Open Questions

In the years just before his death in 1988, the cultural analyst Raymond Williams wrote two books that built directly on his experience of moving from a childhood in the English-Welsh borderlands into a cosmopolitan world of intellectual exchange: the novel *Loyalties* (1985) and an unfinished set of episodes of environmental-historical fiction, *People of the Black Mountains* (1990, 1992). I was led to both these works through an essay by geographer David Harvey, titled "Militant Particularism and Global Ambition" (Harvey 1995). In this essay, Harvey analyzes these works and earlier novels of Williams at the same time as he weaves in reflections on his own experience as a professor at Oxford University.

Harvey had arrived in Oxford in 1989 fresh from completing *The Condition of Postmodernity* (Harvey 1989)—a book that was to become widely read and influential. He was soon drawn in as a "big name" to co-edit a volume of contributions from union activists and academics (Hayter and Harvey 1993). The focus of the project was the decline of the Cowley car plants that had fueled the economy of Oxford since the 1920s—as well as serving as the locus of many significant industrial disputes. During the 1980s, the car plants had been subject to repeated cutbacks. Their closure and redevelopment of the land for a business park was a constant threat. Teresa Hayter, the volume's other editor, had tirelessly campaigned with other shop-floor militants to preserve the plants and the remaining jobs—without the support of the union leadership and local Labour council. When Harvey composed a concluding chapter that entertained other strategies for dealing with the plant closing and the economic future of Cowley, Hayter challenged him to define his "loyalties" (Harvey 1995, 71).

In the "Militant Particularism" essay, Harvey describes his position as wanting to chart a long-term trajectory when, in the short term, there were few alternatives for local workers if the remaining jobs were lost. The situation did not lend itself to a simple reckoning of his loyalties: the working conditions at the plant were deteriorating; the plant's paint shop was a serious source of pollution; working-class solidarities around the plant were weakened and broader coalitions were needed; excess production capacity for cars prevailed in Europe, and indeed, worldwide; Cowley produced Rovers—luxury cars for the wealthy; and the corporate owners of the plant were making decisions based on fluctuations in stock-market and property values. Intellectually, Harvey wanted these issues to be raised so that readers of the volume could "consider active choices across a broad range of possibilities," yet he recognized that "the impetus for the campaign, the research, and the book did not come from [himself, but] arose out of . . . a tradition of union militancy emanating from the plant." He wanted political discussion to be guided by abstractions at spatial and temporal scales larger than the local and immediate, yet felt uneasy—disloyal—about imposing that "upon people who have given their lives and labor over many years in a particular way in a particular place" (Harvey 1995, 73).

Williams's *People of the Black Mountains* resonates strongly with the project of analyzing socio-environmental change in terms of differentiated agents situated in intersecting processes (chapter 5, section C), but it was the novel *Loyalties* that kept me thinking about Harvey's dilemma. Through its central characters, in particular the Welsh Gwyn and his English birthfather Norman, *Loyalties* explores the tension between solidarities forged through working and living together in particular places—"militant particularism"—and translocal perspectives or abstractions. Moreover, it adds a temporal, trans-generational dimension that is especially significant given my interest in "self-conscious knowledge-making and social changing" (chapter 6), or, in Williams' words, in "looking, in [an] active way, at the whole complex of social and natural relationships which is at once our product and our activity" (Williams 1980, 83).

When the middle-aged Gwyn and elderly Norman finally meet, Norman pushes Gwyn to acknowledge that his scientific career has taken him away from his birthplace and enabled him to see more about ways the world is changing than people who remained in the Welsh towns. Political involvement, Norman argues, cannot be a simple matter of Gwyn staying loyal to his roots. Given the "powerful forces" that shape social and environmental change, we can "in intelligence" grapple with them "by such means as we can find" and take a deliberate path of action, but "none of us, at any time, can know enough, can understand enough, to avoid getting much of it wrong" (Williams 1985, 357–58). Or, in the words of Norman's close intellectual and political colleague, Monkey Pitter, if we "go on saying the things we learned to say and it will be just strange talk, in a strange land" (161).

Rather than view these conclusions nihilistically, Williams, in another text that looked toward the future, expressed his hope for "detailed, participat[ory], consciously chosen planning" and opposition to the crisis management and "politics of temporary tactical advantage" he saw ascendant in the 1970s and 1980s (Williams 1983, 11–12). He would, I suspect, have been impressed by what has been achieved in West Nipissing, yet may have expressed uncertainty about its wider implications. To what extent could such local planning mitigate adverse decisions made by governments and corporations operating in a larger spatial and temporal arena? In what ways would it be important to incorporate the knowledge-making of nonlocal or translocal researchers—people who did not share experience of and commitment to livelihood in one place?

When I first read *Loyalties,* I was struck by its resonance with the spirit of my explorations of the construction and reconstruction of unruly complexity. "When inquiries are oriented and guided by themes," the draft introduction of this book manuscript at that time read, "it should be kept in mind that they will break down when applied too far out of the domain in which they were formulated. On the way, however, the explorations enable new terms, questions, models, and relationships to be derived." *Loyalties* has two short passages as "bookends": At the beginning, Jon, a grandson of Norman, is invited to work on a documentary project ostensibly about the politically subversive past in which Norman was implicated. Jon comes to realize that the project is very much connected with political machinations in the present. At the book's end, the final exchange comes after Jon has declined the producer's invitation (Williams 1985, 378): "'As I said at the beginning,' [the producer] shouted, 'you'll cut and run.' Jon stood holding the door. The edge of the wood was between his fingers. 'I told you. I have these questions to ask. Open questions.'"

Summary of Themes and Questions Opened Up

This book considers three angles—like facets of a crystal—from which to view the practice of researchers:

A. their study of complex situations;

B. their interactions with other social agents to establish what counts as knowledge; and

C. their efforts to pursue social change, in which they address self-consciously the complexities of their own situatedness as well as of the complexities of the situation studied.

These angles are identified explicitly in chapter 6, but are evident in the larger structure of the book's three parts. The complex situations referred to in angle A are primarily those studied in ecology and socio-environmental research, but the complexity of influences studied in the interpretation of science leads to an equivalent set of three angles.

For each angle, I discuss problems with simple formulations of well-bounded systems that have coherent internal dynamics and simply mediated relations with their external context (labeled type 1 formulations in chapter 6). I contrast these formulations with work based on dynamics among particular, unequal units or agents whose actions implicate or span a range of social domains (type 3 formulations). I note, however, that

simple formulations are easier to communicate than reconstructions of particular situations and that simple formulations appear to have more effect on
social mobilization. I introduce, therefore, an in-between kind of formulation (type 2): simple themes that open up issues, pointing to greater complexity and to further work needed in particular cases. Indeed, opening out
across boundaries and opening up questions provides the impetus from
each chapter to the next. This mode of expository and conceptual development is conveyed by the following summary of the book's themes and the
questions opened up.

For each chapter, the overall direction is conveyed through themes
(denoted by •) that "point to" or "open up" (denoted by →) a larger project or question (denoted by Q). This direction is also conveyed by the
diagrams, which are subsets of the 3 × 3 framework introduced in
chapter 6. The letters and numbers refer to the angles and formulations
of chapter 6 (see also figure 6.3, which lists specific cases that serve as
examples).

Part I, Modeling Ecological Complexity, considers the use of models to study
complex situations—angle A—but ends by opening up angle B.

Chapter 1, Problems of Boundedness in Modeling Ecological Systems, proposes
that

- the construction of ecological complexity over time, its spatial embeddedness, and the dynamics of unmodeled variables make it problematic
  to theorize about complexity using models of well-bounded systems.
- new concepts, questions, hypotheses, and themes can emerge through
  exploring the qualitative behavior of simple models.

These two themes together open up the questions of

→ Q: how to investigate not only the current configuration of any complex
  ecological situation, but also its particular history and spatial embeddedness
  within intersecting processes (→ A3).
→ Q: how, when using a model heuristically, to assess its limits and minimize
  the chances of applying the model beyond its scope and being misled
  (→ A2).

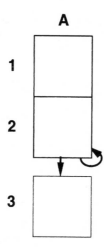

Chapter 2, Open Sites in Model Building, introduces a taxonomy of what ecologists do when they build models in which

- the value of exploratory modeling for theory generation is a counterweight to an emphasis on testing specific hypotheses about particular situations.
- there are always some open sites—categories and relationships accepted without explicit analysis of correspondence with evidence.

It opens up the question of

→ Q: how to identify and make sense of the influences on decisions that modelers make at the open sites (→ B).

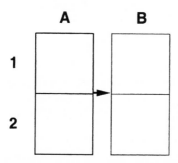

Part II, Interpreting Ecological Modelers in Their Complex Social Context, corresponds to angle B, but, by emphasizing the interpretation of ideas with reference to the actions that those ideas facilitate, opens up angle C.

Chapter 3, Metaphors and Allegory in the Origins of Systems Ecology, interprets the development of H. T. Odum's contributions to system ecology (a case of A1), noting that

- clear correlations can be drawn among Odum's social context, personal experience, and scientific work.
- the mechanism generating these correlations depends on Odum wanting the overlapping domains he inhabited—the social, personal, and scientific—to reinforce one another.

It opens up the questions of

- → Q: how to show reinforcement across domains in cases in which the social-personal-scientific correlations are less obvious or are less consistent over time (→ B3).
- → Q: how to bring such interpretations—ones that show systematic effects of the sociality of ecological science on its referentiality—to bear productively on subsequent research (→ C).

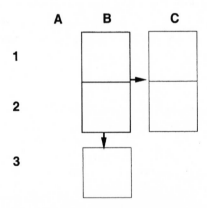

Chapter 4, Reconstructing Heterogeneous Webs in Socio-Environmental Research, interprets the modeling work in two short-term socio-environmental assessment projects (both cases of A1) so as to highlight ways in which

- scientists represent-engage; that is, they establish knowledge and develop their practices through diverse practical choices, such as by mobilizing and connecting diverse resources—that is, by heterogeneous construction.
- the outcomes of scientific work—theories, readings from instruments, collaborations, and so forth—are accepted because they are aspects of heterogeneous webs that are difficult to modify in practice.

- interpretation of scientific work as heterogeneous construction exposes specific points at which concrete alternative resources could be mobilized.

It opens up the question of

→   Q: how to realize the possibility that explicit attention to scientists' diverse resources could help them—or others in comparable situations—alter their personal, scientific, and social facilitations, and so modify the directions in which their science moves (→ C).

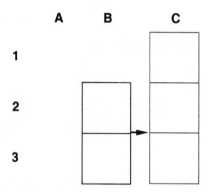

Part III, Engaging Reflexively within Ecological, Scientific, and Social Complexity, corresponds to formulations 1 and 2 of angle C, but ends by opening up formulation 3 of angle C for future work.*

Chapter 5, Reflecting on Researchers' Diverse Resources, reviews my efforts to engage researchers—interpreters of science and scientists—in analyzing researchers' diverse resources, and proposes that

- interpreting science as heterogeneous construction requires conceptual and methodological choices in which practical considerations are implicated, which means that interpretation also involves heterogeneous construction.

---

*Part III also includes some reflexive interpretation of research that interprets science. If the object of research is the complexity of influences on the practice of scientists, then angles A′, B′, C′ on the practice of researchers can be defined, where "researchers" refers now to the interpreters of science.

- interpreters of research as heterogeneous construction should distribute the
  work of interpreting and engaging with that research, for example, by
  leading researchers to map the situations they study and their own
  situatedness, or by stimulating them to take initiative in mobilizing new
  resources that support new directions in their work.
- there is a tension between the logic of exposing the situatedness of
  particular researchers—scientists and interpreters of science—and pragmatic
  choices that limit the probing of conceptual and methodological choices
  and keep situatedness in the background.

It opens up the question of

→ Q: how, in practice, to open up researchers' situatedness in ways that
   facilitate its reconstruction (→ C).

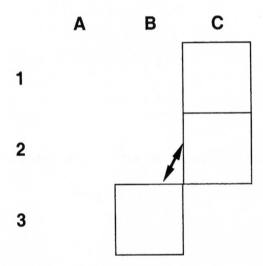

Chapter 6, Reasoned Understandings and Social Change in Research on Common
Resources: Introducing a Framework to Keep Tensions Active, Productive, and Ever-
Present, acknowledges that

- system-like formulations are easier for members of an audience to digest
  than are particular cases of intersecting processes, heterogeneous
  construction, or practical reflexivity, but

- it is possible to introduce simple themes that open up issues, pointing to greater complexity and to further work needed in particular cases. Indeed,
- knowledge-making agents are always moving:

    a. between system-like formulations and accounts of unruly complexity;
    b. among three angles for viewing their own practice—dialogue with the situation studied, interactions with other social agents to establish what counts as knowledge, and efforts to pursue social change by addressing the complexities of their own social situatedness as well as the complexities of the situations they study; and
    c. between a concentrated view of their agency and awareness of conditions for modifying or restructuring their situatedness that are more distributed and dependent on the actions of other agents.

It opens up the question of

→ Q: how individuals, with their knowledge, themes, and other awareness of complex situations and situatedness, can participate with others in restructuring the distributed conditions of knowledge-making and social change (→ C3).

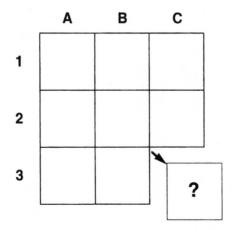

Glossary

This glossary is included to remind readers of the specific or technical meaning attached to key terms and thus to reduce the chance that readers will dismiss unfamiliar terms as unnecessary jargon. The places where the terms are introduced or elaborated on are given in parentheses. Italicized terms within definitions are those defined elsewhere in the glossary.

Accessory conditions. Features that are built in and assumed when analyzing a *model* but which are not the distinguishing features that explicitly characterize that model (chapter 2).

Adaptive Environmental Management (AEM). Environmental management that is adjusted (along with the models used in the formulation of management policy) following assessment of the outcomes (such as the outcomes of deliberate policy experimentation). AEM begins with recognition that the dynamics of any ecological *situation* are not fully captured by any *model* or composite of models, especially because management practices produce continuing changes in those dynamics and make the ecological situation a moving target (note 19).

Agent. A person whose thinking, decisions, and actions the analyst or interpreter is trying to associate with specific environmental, scientific, or social changes (chapter 3ff). See also *concentrated, distributed,* and *vibrating agency.*

Apparent interactions. The values of parameters in a *model* that best fit the observations of part of an *ecological community* when that model makes no explicit reference to the dynamics of the rest of the community. These values are not necessarily close to the direct interactions and may be counterintuitive (chapter 1, section B).

Community. See *ecological community.*

Complexity. A composite of the number, interconnectedness, heterogeneity, and variability (in space and time) of the factors influencing the *situation* under study (chapter 1, section A). See also *unruly complexity.*

Concentrated agency. Thinking, decisions, and actions of *agents* that emphasize their independence from other agents and their ability to impart order according to their ideas or beliefs (chapter 4, section A). See also *distributed* and *vibrating agency.*

Conceptual exploration. Playing with *themes* and *models* so as to open up questions in broad terms that might transfer across different fields while, at the same time, keeping the limitations of such themes and models in view (prologue).

Confirmation. Showing that the distinguishing features of a causal *model* and, independently, the model's *accessory conditions* correspond to observations.

Construction. Processes through which entities are connected over time (or through definite steps) into an outcome that has *structure* and is subject to ongoing restructuring (chapter 1, section A; chapter 4). See also *heterogeneous construction.*

Counterfactual. An idea of what else could have been if some feature of a *situation* were replaced by something else—things that might have occurred but did not. The counterfactual is logically implied whenever there is a causal claim that the feature made a difference (chapter 4, section A; notes 17 and 20).

Critical reflection on practice. Self-conscious, systematic efforts by researchers or other *agents* to understand the social and technical conditions in which their practice takes shape by placing those conditions in tension with alternatives (prologue).

Distributed agency. Thinking, decisions, and actions of *agents* that recognize their contingent and ongoing mobilizing of *webs of resources* (diverse materials, tools, and other people) (chapter 4, section A). See *heterogeneous construction; concentrated* and *vibrating agency.*

Ecological community. A defined group of interacting populations of different species, which forms part of an *ecosystem* (narrative at the start of part I; note 5).

Ecological-like complexity. Unruly complexity in realms other than ecology (prologue).

Ecological system or *ecosystem.* A defined set of living and nonliving entities (or "compartments") among which energy and nutrients flow (narrative at the start of part I; note 5).

Ecology. (1) The diversity of interactions that some organisms or people are involved in (whether directly or indirectly through extended chains of influence). The *complexity* of those interactions is *constructed* over time. (2) Scientific research or knowledge regarding those interactions. (3) Social actions responding to the degradation of the environment of humans and other species (prologue; chapter 1ff).

Embeddedness. The interconnectedness of the dynamics of a *situation* and the dynamics of another situation at a larger scale in time and space (prologue). See also *situatedness* and *intersecting processes.*

Engagement. An *agent*'s deliberate involvement in a *situation* in ways that presume that other agents will also take an active role and that the situation cannot be understood or managed from an outside vantage point. Engaging implies a more participatory spirit than intervening (chapter 4ff). See also *representing-engaging.*

Exploratory modeling and *theorizing. Conceptual exploration* that uses *models* in the sense of *frameworks* to derive new terms, questions, and theoretical relationships. Contrasts with theorizing that is centered around specific hypotheses that can be readily tested or, more generally, that favors making incremental changes to the current terms, tools, and subject matter of a field (chapter 2).

Flexible engagement. An ideal in which researchers in any knowledge-making situation are able to connect quickly with others who are almost ready—either formally or otherwise—to foster participatory *processes* and, through the experience such processes provide their participants, contribute to enhancing the capacity of others to do likewise (epilogue, section B).

Framework. One or more propositions or *themes* that guide the inquiry of researchers by focusing their attention on certain *situations* and *processes* (chapter 2).

Heterogeneous construction. Construction of science-in-the-making from diverse materials, tools, people, and other *resources.* In heterogeneous construction, researchers establish knowledge and develop their practices through diverse and often modest practical choices, which is the same as saying they are involved in contingent and ongoing mobilizing of materials, tools, people, and other *resources* into *webs* of interconnected *resources* (chapter 4, especially the end).

Heterogeneous constructionism. The perspective that science-in-the-making can be interpreted as *heterogeneous construction* (chapter 4).

Heterogeneous web. See *web of resources.*

Interpretive studies of science. Historical, philosophical, and sociological studies concerned with explaining or describing the dynamics and course of scientific inquiries (narrative at the start of part II).

Intersecting processes. Processes operating at different spatial and temporal scales that transgress the boundaries of the *situation* under consideration and restructure its "internal" dynamics. The term characterizes the same terrain as *unruly complexity,* but is used to suggest that different strands of the processes can be teased out in a somewhat disciplined fashion (narrative at the end of chapter 1; chapter 5, section C). Viewing the activities of researchers in terms of intersecting processes is very similar to *heterogeneous constructionism.*

Mapping. A process in which researchers identify "connections" to a key issue—things that motivate, facilitate, or constrain their inquiry and action—then create and revise pictorial depictions—maps—that employ conventions of size, spatial arrangement, and perhaps color so as to allow many connections to be viewed simultaneously. The map *metaphor* connotes not a scaled-down *representation* of reality, but a device that shows the way—a guide for further inquiry or action (chapter 5, section C).

Metaphors. Features and associations borrowed or carried from one *situation* so as to animate our thinking about another situation. This process unavoidably affects our thinking about the primary situation. Identifying scientists' metaphors can illuminate why certain categories are plausible and certain lines of inquiry are pursued (chapter 3 and note 11).

Models. Mathematical equations, verbal or pictorial *representations*, or material devices that researchers use to depict certain *situations* and examine the relationships among certain entities in those situations. The status or use of a model depends on the level of correspondence with observations (the degree of fit and the strictness with which *accessory conditions* have been established) and the ways in which the modeler attempts to expand or disturb acceptance of the model (chapter 2). See also *theory.*

Nomadic pastoralists. Herders living in semiarid climates where rainfall is variable, unpredictable, and spatially patchy, who spend at least part of their year roaming in search of patches of watered pasture. Within this definition, the historical and geographic variation is enormous (chapter 4, section B).

Opening-up themes. Themes that build in a persistent opening up of issues—that point to the hidden *complexity* of simple formulations and to further work needed to address the differentiated detail and other complexities of particular cases (chapter 6).

Organicism. The perspective that social or ecological organization consists of components that work together in a hierarchical division of labor to maintain the unity and *stability* of the whole, just as the organs of an organism do. Terms such as "adjustment," "adaptation," "integration," and "function" are common in organicist discourse (chapter 3).

Participatory action research (PAR). Inquiry that is shaped by researchers' ongoing work with and empowerment of the people most affected by some aspect of economic or social change (note 31).

Partitionability. The property of two *situations* whereby their dynamics are at most loosely linked to each other; the opposite of one of the situations being *embedded* in the other or of a situation being affected by *intersecting processes* (prologue; see also note 17).

Political ecology. The study of specific cases of environmental degradation in terms of links among local changes in agroecologies, labor supply and the organization of production, and wider political-economic conditions (narrative at the end of chapter 4).

Practical reflexivity. Reflexivity that takes into account the range of practical conditions that enable researchers to build and gain support for their *representations* (chapter 5, section A).

Processes. Sequences of events that persist or are repeated long enough for us to notice them and need to explain them. In this book, *process* is not used in the sense of a basic underlying causal structure that allows people to explain events as instances of the process or as noisy deviations from it (note 23).

Referentiality. The relationship of a *model* or *theory* to scientific observations of the phenomenon modeled (narrative at the start of part II).

Reflexivity. The application of one's analytic approach to one's own work (note 21).

Representation. A *model* or *theory* used to depict certain *situations* and draw attention to relationships among certain entities in those situations (chapter 2).

Representing-engaging. The process of deriving a *representation* in which scientists are simultaneously and jointly acting or *engaging* in various arenas of social activity—building careers and institutions, using and transforming language and ideology, facilitating policy formulation, and so on (chapter 4).

Resources. Materials, tools, persons, and other things mobilized by researchers during the *heterogeneous construction* of science that make a claim or a course of action more difficult for others to modify. By extension, a resource for one person is a constraint for another person trying to modify the first's claim or action (chapter 4).

Situatedness. Embeddedness of *agents* in social *situations*. Situated scientific agents are mindful of both the situation they are studying and the social realms in which they act, and that they project continuously between the situation studied and their social situation (chapter 4).

Situation. A catchall term that, like *system*, denotes that the phenomena under consideration have many interacting elements but, unlike *system*, does not connote boundedness, coherent dynamics, or *stability* (prologue).

Social construction. A *process* by which scientists weave their social context or their experience of being social *agents* into the knowledge they establish (note 14). See also *representing-engaging* and *heterogeneous construction.*

Sociality. Interactions among researchers to establish what counts as knowledge, involving disputes and dialogue around methods, observations, conclusions, and practical applications (narrative at the start of part II).

Social-personal-scientific correlation. The interpretation of the outcomes of research as supporting and supported by researchers' concerns about social order and their personal life histories (chapter 3).

Socio-environmental studies. Research in fields such as environmental science, resource economics, geography, and anthropology that focuses on the ways social relations influence environmental change and the dynamics of ecological situations influence social change (chapter 4).

Stability. The degree to which a *system* resists perturbations in its composition or in the sizes of its components and returns to its former state after perturbation—in particular, to a steady state or equilibrium point (chapter 1, section A).

Strong system. See *system.*

Structure. Regularities in *situations* that persist long enough for *agents* to recognize or abide by them (prologue; chapter 5, section C).

System. (1) A phenomenon in which there are many elements interacting (loose view). (2) A well-bounded entity in which many components interact and which has coherent internal dynamics that govern the system's development and responses to external influences (strong view) (prologue; note 5; chapter 3; chapter 4, section B; chapter 5, section C; chapter 6).

Systems ecology. The scientific study of the flow of nutrients and energy through an entire *ecosystem,* including the decomposer components (chapter 3).

Technocratic management. (1) The use of technical measures to address social problems (loose view). (2) Through the use of science to reduce social complexity to mechanical relationships, scientists and engineers manage society at a distance from specific interests and political details and in the best interests of everyone (strong view). As is typical of social philosophies framed in terms of universal interests, proponents of technocratic management build a special place for themselves in the proposed social organization or governance (chapter 3; chapter 4, section B).

Technocratic optimism. Optimism after World War II about the possibilities for *technocratic management* of society, which left little room for doubts about its possible implications for democratic political life (chapter 3).

Theme. A catchall term that denotes propositions, concepts, analogies, *models,* or heuristics that researchers use to focus attention on certain *situations* and *processes* when the intent is not immediate empirical validation, but to stimulate our thinking, open up questions, and orient our inquiries. In this spirit, themes should be expected to break down when applied broadly (chapter 2). See *conceptual exploration.*

Theory. One or more *models* making up a "fabric" such that only some models—those at the edge of the fabric—are exposed for empirical scrutiny. Other models woven into the theoretical fabric may be left inexplicit (chapter 2).

Trajectory. Changes over time in the size of populations in an ecological community.

Unruly complexity. Situations that do not have clearly defined boundaries, coherent internal dynamics, and simply mediated relations with their external context—that is, are not strong *systems.* Instead, there is an ongoing change in the *structure* of the situation as it builds up over time from heterogeneous components and is *embedded* or *situated* within wider dynamics (prologue; chapter 5, section C).

Vibrating agency. Thinking, decisions, and actions of *agents* that move between *concentrated* and *distributed* views of their own *agency.* Although *system*-like formulations about one's social *situatedness* serve as *resources* for agents in their knowledge-making, the same agents can acknowledge their dependency on diverse materials, tools, and other people that are mobilized in particular, contingent social *situations* (chapter 6, section C).

Web of resources. Diverse *resources* that are interconnected in the practice of science (chapter 4). See also *heterogeneous construction.*

Notes

The main body of the text is written to engage a range of audiences, so I minimize specialist discussion that is not essential for the flow and substance of the text. The thematic notes that follow are included to elaborate on some technical issues and specialized debates, and on developments subsequent to the period in which the cases originated. In these notes I also acknowledge some related projects and position my approach with respect to them. My intent is to stimulate dialogue, not to provide a comprehensive bibliography or overview of any issue.

Prologue

1. The critique of science during the 1970s

During the 1960s, Bookchin (1962), Carson (1962), and Commoner (1963, 1971) linked ecology-as-social-action to criticisms of the dominant directions of scientific research. Social responsibility in science was promoted by the British Society for Social Responsibility in Science, the Union of Concerned Scientists (in the United States), and the Pugwash Conferences on Science and World Affairs (which focused on reducing the danger of armed conflict in a nuclear age). During the 1970s, more radical, anticapitalist critiques of science and social relations were developed in the United States by Science for the People (see especially the critiques of biological determinism in Science for the People 1977 and the related work of Chase 1977) and in England by the Radical Science Journal (see Levidow 1986, especially Young's introduction on the origins of the "Radical Science" movement; Levidow and Young 1981; Radical Science Editorial Collective 1977). In this context, Werskey completed his illuminating history of an earlier generation of left-wing scientists in England who saw "science, progress, socialism as equivalent concepts" (Young, p. xiv in the foreword to the 1988 reprint of Werskey 1978; see also Werskey's preface to the reprint, which reflects on the way the 1970s critique of science shaped his account). In contrast, left-wing scientists of the 1970s who saw their science as a political project recognized that science could bolster domination and inequality (Roberts 1979; Rose 1982; Levins and Lewontin 1985; see also Illich 1973 and 1976 for advocacy of deprofessionalization and of "convivial" technology and medicine).

The critique of science also stimulated interpretation of science in relation to the historical and social context in which it was formed. Kuhn (1970) was widely cited as opening up science to such contextualization, but younger historians and sociologists began to take the social interpretation of science much further than Kuhn had (see, for example, Young 1985).

2. Intellectual trajectories away from comprehensive quantitatively based analysis and planning

Many researchers who during the 1960s and 1970s were interested in comprehensive, quantitatively based analysis and planning have since left this ambition behind. Based on personal conversations, I believe that a "Modelers Anonymous" organization would have many potential members! David Harvey, a geographer who is featured in section C of the epilogue, might be one example (compare Harvey 1969 with Harvey 1995; see also autobiographical essays by Harvey and by Olsson in Gould and Pitts 2002).

3. Learning about ecological and scientific complexity through long-term engagement in particular settings or in larger social mobilizations

Examples of texts that address the intersections of environmental and social change through long-term engagement in particular settings or in larger social mobilizations, and which thus stand in potentially productive tension with this book's conceptual exploration, include Boal (2005); Brown and Mikkelsen (1990); Gibbs (1982); Hofrichter (2000); Martin (1996a); Pellow (2002); Sclove (1995); and Williams (1990, 1992). Some relevant organizations in the United States include the Community Research Network (associated with the Loka Institute, http://www.loka.org/), the Highlander Education and Research Center (http://www.highlandercenter.org/), and the Center for Health, Environment and Justice (formerly Citizens Clearinghouse for Toxic Wastes, http://www.chej.org/). See also the epilogue.

4. Studies of complexity in ecology, socio-environmental studies, and interpretation of science

In addition to the work identified in the previous note, I would place the following texts on my provisional map of positions that complement or stretch this book's approach to complexity in ecology, socio-environmental studies, and interpretation of science:

Ecology: Haila and Levins (1992), Levin (2000), Pimm (1991).
Socio-environmental studies: Berkes, Colding, and Folke (2002), Guyer (1988), Peet and Watts (1996a), Roe (1998), Turner and Taylor (2003).
History of ecology: Bocking (1997), Hagen (1992), Kingsland (1995), Kwa (1989).
Social studies of science, especially complexity and modeling: Edwards (2005), Fausto-Sterling (2000), Gieryn (1999), Haraway (1988), Helmreich (1999), Latour (1987, 2004), Law and Mol (2002), Pickering (1995), Traweek (1994). (In Law and Mol 2002 see especially the editors' introduction and essays on ecological themes by Kwa and Thompson.)
Philosophy and conceptual studies of science: Bohm (1995), Krieger (1994), Longino (1990), Oyama, Griffiths, and Gray (2001), Wimsatt (2001). (In this area there is room for dialogue around epistemology construed broadly as

inquiry about what makes it possible for social agents to make reliable knowledge; see also notes 17 and 20.)

Studies of gender or race in relation to complexity: Eglash (1999) and the works cited above by Fausto-Sterling, Guyer, Haraway, Krieger, Law and Mol, and Traweek. (These are important directions of stretching given that almost all the researchers in the book's cases are men of European descent, as I am also; but see note 28.)

Critical reflection on practice: See the epilogue and notes 29–31.

Part I

5. Theorizing about ecological complexity through the mid-1980s

A broad distinction can be made between community ecology, which emphasizes population sizes and interspecific interactions, and systems ecology, which emphasizes nutrient and energy flows between ecosystem compartments (Hagen 1989). Nevertheless, community ecological theory also involves systems in the sense of entities that have clearly defined boundaries, coherent internal dynamics, and simply mediated relations with their external context (Taylor 1992a, 2001c, 383ff). (See also the synthesis of the two schools in DeAngelis 1992 and Taylor and Post 1985; the latter relates specifically to the complexity-stability relationship discussed in chapter 1, section A.) One needs to go beyond this dichotomy, however, to capture the range of basic impulses in studying ecological complexity evident in United States ecology from the 1950s through the mid-1980s. I identify these impulses in the table below, with position from left to right used to denote earlier or later emergence:

BASIC IMPULSES in the study of ECOLOGICAL COMPLEXITY

---

1. Ecosystems are complex, yet have SYSTEMIC PROPERTIES

    2. Ecosystems are complex systems of COMPARTMENTS and FLOWS of energy and nutrients

        3. Ecological complexity will be built up from BASIC, GENERAL RULES, especially about populations and their interactions

            4. HIERARCHY THEORY—find natural scales of patterns and processes

        5. Important influences will be evident in PATTERNS of ecological complexity, as revealed by diversity measures or other community descriptors, or through multivariate analyses

            6. Rules and generalizations may emerge from attention to actual PROCESSES through experimental manipulations or long-term observations

                7. PARTICULARISM—No generalizations from situation to situation

            8. Ecology advances by refutation of TESTABLE HYPOTHESES

Chapter 1

6. Revival of interest since the mid-1990s in explaining ecological complexity

Since the mid-1990s, there has been some revival of interest in explaining diversity
(Tilman and Pacala 1993) and its contribution to ecosystem dynamics. Experiments and
models support the idea that stability, in the sense of constancy of numbers, increases
with the number of species in mixtures of plants (Tilman 1999). Yet productivity, more
than stability, is the attribute that plant ecologists now try to associate with diversity,
and even there, the significance of results in this area is hotly disputed (Kaiser 2000b;
Huston et al. 2000). In any case, this research focuses on the single trophic level of
plants, not the complexity of whole communities or ecosystems.

7. Construction as a metaphor in biology

To use the term *construction* in a biological context requires one to deemphasize
connotations of an intentional agent doing the construction—a problem shared with
the related terms *self-organization* and *assembly* (Taylor 1998c; Odling-Smee, Laland,
and Feldman 2003, 41n). This choice of term allows me to draw parallels later in the
book with approaches in other fields that address heterogeneity of components,
ongoing restructuring, and embeddedness in wider dynamics. See note 14.

8. Theorizing about ecological complexity since the mid-1980s

The constructionist and landscape views reinforce other currents that have undermined the aspirations of earlier decades to identify general principles about systems
and communities (Kingsland 1995, 213–51; Taylor and Haila 2001). Since the 1980s,
ecologists in general have become increasingly aware that situations may vary according to the historical trajectories that have led to them; that particularities of place and
connections among places matter; that time and place are a matter of scales that differ
among co-occurring species; that variation among individuals can qualitatively alter
the ecological process; that this variation is a result of ongoing differentiation occurring within populations—which are specifically located and interconnected—and that
interactions among the species under study can be artifacts of the indirect effects of
other "hidden" species (see chapter 1, section B).

In patch dynamic studies, for example, the scale and frequency of disturbances
that create open "patches" are now emphasized as much as species interactions in the
periods between disturbances (Pickett and White 1985). Studies of succession and of
the immigration and extinction dynamics of habitat patches pay attention to the particulars of species dispersal and the habitat being colonized, and to how these determine
successful colonization for different species (Gray, Crawley, and Edwards 1987).
Metapopulation theory examines the persistence not of communities, but of populations (or phoretic associations of communities on carrier species) in a landscape of
patches (Hastings and Harrison 1994). On a larger scale, such a shift in focus is supported by biogeographic comparisons showing that continental floras and faunas are
not necessarily in equilibrium with the extant environmental conditions (Haila and
Järvinen 1990). From a different angle, models that distinguish among individual
organisms (in their characteristics and spatial location) have been shown to generate
certain observed ecological patterns, such as patterns of change in the size distribution

of individuals in a population over time, where large-scale, aggregated models have not (DeAngelis and Gross 1992). In addition, the effects mediated through the populations not immediately in focus or unrecognized upset the methodology of observing the direct interactions among populations and confound many principles, such as the competitive exclusion principle, derived on that basis (chapter 1, section B; Wootton 1994).

By and large, philosophers of ecology, environmental ethicists, environmentalists, and others who invoke principles of ecology have yet to address the implications for their fields of this picture of ecological complexity (Taylor 1997b; Taylor and Haila 2001).

9. Apparent interactions: A comparison of alternative derivations

There are many references to apparent interactions or indirect effects in the ecological literature (e.g., Levine 1976; Holt 1977; Lawlor 1979; Vandermeer 1980; Schaffer 1981; Bender, Case, and Gilpin 1984). To understand their differences, it is first necessary to distinguish several ways of defining and estimating interactions from data.

Interactions: Definitions and estimation

Interactions may be directly observed, but these are interactions between individual organisms. Alternatively, interactions between populations may be inferred from data. This requires either a theory that tightly links a qualitative outcome (e.g., non-coexistence of similar populations) with the interaction (e.g., competition) or quantitative data on population sizes. Two types of interactions may be derived from such quantitative data:

1. The data may be construed as population sizes changing over time under a fixed set of conditions. The effect of population j on population i is then defined as its contribution to altering the rate of change of i. There are three major approaches to the estimation of such effects:

   a. Fit the data to a model postulated to govern the interacting populations; that is, estimate the values of the parameters for which the model best fits the data.

   b. Infer interactions from data on experimental perturbations from equilibrium. Bender, Case, and Gilpin (1984) propose two procedures, "PULSE" and "PRESS," for estimating the parameters of the generalized Lotka-Volterra (GLV) model (model 1 in the text). These procedures require that the populations are initially at equilibrium and that a controlled series of perturbations from that equilibrium can be performed.

   c. Infer interactions from data on moving equilibria.

To understand the last, somewhat paradoxical, approach requires the introduction of the second way to derive interactions from quantitative data on population sizes:

2. The data may be construed as moving equilibria of the population sizes under conditions that are changing, perhaps as induced experimentally.

The effect of population j on population i can be defined in terms of the covariation of their equilibrium values—does one go up when the other goes down, or do they go up and down together? Such interactions between populations are usually limited to competition, mutualism, or null interactions. This concept of interaction is not closely related to either directly observed interactions or those derived from data construed to be population sizes changing over time under a fixed set of conditions (type 1 interactions). If values for type 1 interactions (approaches a or b) are estimated, then type 2 interactions can be calculated by loop analysis (Puccia and Levins 1985). If values for type 2 interactions are estimated, then the qualitative values of type 1 interactions can sometimes be inferred by inverse loop analysis (Puccia and Levins 1985)—this constitutes approach c above.

Apparent interactions: Definitions and estimation

Apparent type 1 interactions. Apparent type 1 interactions should generate trajectories for the members of an apparent community that mimic the actual trajectories for those members (i.e., the trajectories generated by the dynamics of the full community). Variations of this approach depend on whether the interactions

a.   are assumed to govern trajectories near equilibrium—the method used in the text;
b.   are calculated as if the community were near equilibrium, but then extended to apply away from equilibrium by using a GLV (model 1 in the text);
c.   are derived by fitting observed trajectories directly to the GLV model; or
d.   are derived by fitting observed trajectories directly to some model other than the GLV.

Other methods include a generalization of MacArthur's (1972, 33ff) and Schaffer's (1981) Abstracted Growth Equations. Assuming certain special conditions, method d can be used to derive equations for the apparent community when we do not have knowledge of the full system.

Bender, Case, and Gilpin's PULSE and PRESS methods. If the timescales of the hidden populations and of the modeled populations are similar, the PULSE method yields estimates of the direct interactions between the modeled populations. These estimates are unable, in general, to mimic the actual trajectories of the modeled populations (see the text). If the timescales are disjunct, then PULSE estimates of the direct interactions may absorb effects from the hidden variables. In other words, they will be estimates of apparent interactions and thus potentially take counterintuitive values.

The PRESS method has some significant limitations. It cannot be used in most cases in which direct self-interactions are zero (especially those of hidden variables), or where the apparent community has only one member. When the PRESS method can be applied, the estimated interactions are actually estimates of apparent interactions.

Type 2 interactions. Type 2 interactions focus only on the two populations in question and so, in principle, are not affected by the dynamics of hidden variables. However, if loop analysis is used to calculate the values of type 2 interactions from estimates of type 1 interactions, then the full set of direct interactions must be known. (Loop analysis using apparent type 1 interactions of the form developed in the body of the text generate qualitatively good estimates of type 2 interactions, but calculation of such apparent interactions requires knowledge of the full set of direct interactions: Taylor 1985, 119–77.)

Chapter 2

10. Pattern and process: Challenges in ecological data analysis

Multivariate statistical (or data analytic) techniques have long been descriptively used, especially in vegetation ecology, to cluster ecological sites into distinct communities (classification) or position them along continuous axes (ordination). The patterns exposed have also been used to generate hypotheses about causal factors or underlying environmental gradients. The results of pattern analyses, however, are sensitive to the models underlying the technique used and the sampling from the space of environmental possibilities (Faith, Minchin, and Belbin 1987; Minchin 1987). Popular techniques, such as principal components analysis and detrended correspondence analysis, when tested on simulated data, do not recover the simulated environmental gradients well. Techniques that reduce this model dependence also tend to produce degenerate patterns (Faith, Minchin, and Belbin 1987). The catch-22 is that one needs to know a lot about the causal factors behind the data in order to design efficient and distortion-free multivariate techniques that would expose those factors (Austin 1980, 1987).

To some extent, such problems can be overcome through the use of analysis of variance and related statistical techniques on data from replicated, multifactorial field experiments (Underwood 1997). Strictly speaking, however, such results are local—that is, contingent on the configuration of other factors held experimentally or statistically constant for the experiment (Lewontin 1974). Localization poses few problems when ecological engineering affords control over conditions and isolates the system from any surrounding dynamics. But these are special cases; in naturally variable situations, observations constructed for testing of specific, single-factor hypotheses may not be useful for thinking about anything beyond the local configuration observed. Similarly, lack of generality (Kelt and Brown 1999, 99) seems also to be the case for assembly rules in community ecology (Weiher and Keddy 1999). In the absence of information about historical trajectories, *assembly* rules are better thought of as *patterns* of co-occurrence that are statistically significantly different from patterns that are produced by randomly sampling ("assembling") species from the appropriately delimited species pool (Kelt and Brown 1999).

Lying between local results and pattern analysis is the pragmatic use of statistical techniques: if the statistical analysis shows a significant pattern, then ecologists hope it is worth their while to investigate what could be causing that pattern. The risks, however, are that statistical terms—effect, source of variation, and so forth—will be read in causal terms and that alternative categories for data collection will not be considered.

These same issues of pattern and process apply to landscape ecology (Klopateck and Gardner 1999) and geography (Olsson 2002), including studies that make extensive use of remote sensing and Geographic Information Systems (see papers in Turner and Taylor 2003 for critical analyses by practitioners).

Chapter 3

11. Metaphors in science

Philosophers and other interpreters of science have long been interested in the use of metaphors (Benjamin, Cantor, and Christie 1987; Ortony 1993; Taylor 1996a). The general idea, advocated in particular by Hesse (1966), is that analogies, metaphors, and models are alike in allowing the associations from one field to animate a scientist's thinking about another field. Black (1962) observed that this animation comes to work both ways. For example, Darwin's metaphor of natural selection originally invited readers to think about the survival and reproduction of organisms with useful variations as if nature were selecting those organisms. More than a century of use of this metaphor in science has contributed to the sense that when selection is intentionally performed, say, by educational testers, the ranking follows a principle given by nature, not merely a social convention.

Metaphorical associations carried from one field to another are *open* in three senses: they are not fully explicated; they vary among users and readers; and they vary for the same user among different contexts. Notwithstanding the openness of the associations, analogies have limits, shaping the thinking of scientists so that it fits within certain frameworks. Unlike Hesse, however, I do not distinguish positive, negative, and neutral aspects of analogies. As chapter 3 shows, what others consider in hindsight to be limits of an analogy actually facilitated Odum's work, making the scientist's research "do-able" (Fujimura 1987) given the context of the scientific institutions and wider social sentiments of his time.

In the analysis of the use of metaphors I see three related "meta-metaphors" of likening—of how associations are carried from one field to another:

1. Root, fundamental, underlying things shape the surface layers or visible forms
2. Mental things—thoughts, expectations, images we have seen—shape our actions
3. Culture or society gets into these thoughts, expectations, and images (and thus we can be taught how to perceive the world)

These meta-metaphors are not conducive to the idea, developed in chapter 3 and subsequent chapters, that action and thought are jointly shaped through *practical activity* in which a diversity of resources are employed. Although thinking, imaging, viewing, speaking, and writing are actions, they are actions of particular kinds; "acting as if" can be viewed as a more inclusive meta-metaphor of likening. For relevant examples of these dominant meta-metaphors, see Lakoff and Johnson (1980); Stepan (1986); Danziger (1990); Gergen (1990); Lakoff (1993); and Reddy (1993). Insight into the difficulty of moving beyond mental and verbal construals both of metaphor and of action is provided by Mitchell's (1990) discussion of the distinction between persuading and coercing as a master metaphor in social theory (Taylor 1997a, 209–10).

12. Interpreting social order and the twentieth-century life sciences

A number of interpreters of science have interpreted episodes in twentieth-century life and social sciences in terms of concerns about social order and disorder (e.g., Haraway [1981–1982, 1983, 1984–1985]; Cross and Albury [1987]; Gilbert [1988]; Mitman [1994]). Social concerns change as U.S. society changes, and social change is, in part, conditioned by changes in those sciences. Such interpretations extend Williams's (1980) germinal work interpreting ideas of nature as projections into nature of people's ideas about the social order they favor. The interpretations show that society has been naturalized and nature socialized not only in popular ideas about nature, but also in the sciences of nature themselves. Of course, reciprocal processes of naturalization and socialization occur in unevenly changing and partial ways, which are sometimes contradictory. For example, organismic metaphors in social and biological thought gave way after World War II to both cybernetic and individualistic metaphors (Haraway 1981–1982; Mitman 1994; Keller 1988; see also Moore 1997 for an entry point into an analogous literature interpreting Victorian science in terms of a "common context"—or fragmenting common context—of politics, economics, demographics, and institutional developments).

Where connections between science and social order are observed, the question of agency arises: how did scientists and allied agents do their work in a way that one can later interpret as corresponding to concerns about social order? Haraway argues that it is not possible for biologists to do otherwise: like all organisms, they are "material-semiotic" beings, so that "understanding the world [or "worldly practice"] is about living inside stories" (Haraway and Goodeve 2000, 107). Because the embeddings are multiple, the interpreter of science needs to look for "diffraction patterns [which] record the history of interaction, interference, reinforcement, difference" (102; see also Law and Mol 2002). But notice here the reliance on "mental or verbal images, images that we *believe* or *think* that the world is like, or that we *speak* or *write* as if it were like" (Taylor 1997a, 209). In contrast, the interpretive approach developed in chapters 3 and 4 relies less on images of mind-based subjectivity (through which social ideas can be projected more or less directly into scientific ideas) and attempts to make more space for examining the material aspects of specific scientific practice (see note 11).

13. Pictorial representation

By the late 1980s, the analysis of pictorial, graphic, or visual representation in scientific practice had begun to be actively taken up by sociologists of science, with historians and philosophers following in the 1990s. See the collections edited by Lynch and Woolgar (1990), Ruse and Taylor (1991), and Jones and Galison (1998); the introductory essays of these collections; and the references cited therein. See also

Rudwick (1976) for a significant earlier contribution in an area of modern science and Mitman (1999) for an account of nature represented in the medium of film.

This work on pictorial representation complements and sometimes challenges the traditional emphasis of interpreters of science on text and mathematical formulations. As Lynch observes, the process of establishing scientific results almost always requires making something visible and analyzable. Therefore, investigation of representational or communicational devices and practices has a lot to contribute to discussions of "rationality, experimental procedure, observation, and representation" (Lynch 1991). It should be noted that "representation" in this context denotes the process of representing as well as the products, although it is the latter that are the subject of chapter 3, section B. A diversity of currents are evident in the growing literature (Lynch 2001 and references therein), and unresolved questions remain—not surprisingly, given that discussions of perceptual, conceptual, verbal, and graphic images are inextricably cross-referential (Mitchell 1986).

Chapter 4

14. Constructionist interpretations of science

In its broadest sense, the claim that scientific knowledge is constructed amounts to saying that it is not simply drawn from nature. According to a minimal definition of constructionism, what counts as knowledge is contingent on the scientific method or framework used. A more inclusive, *social* constructionism views knowledge as contingent on the scientists establishing (or disputing) the knowledge and then, through them, on their social context (see Collins's [1981b] "stages in the empirical programme of relativism.") During the 1980s, social constructionism (or constructivism) became the major perspective in social studies of science. Active debate ensued about different interpretations of construction, the degree to which it is a social process, how much that process affects what counts as knowledge in the long run, the emphasis on knowledge alone or as integrated into scientific activity more generally, and whether social studies of science now needed to move beyond constructionism. See, for example, the exchanges among Collins, Yearley, Latour, and Woolgar in Pickering (1992b) and between Sismondo (1993a, 1993b) and Knorr-Cetina (1993); Smith's (1997) discussion of Kitcher (1993); and other exchanges reviewed in Hacking (1999).

The sense of construction as building—the process, not the product—remained underdeveloped through these debates (as noted also in Hacking 1999, 49ff). If the project of interpreting science is construed very broadly as addressing what it means *practically* for *agents* to *modify* scientific *activity*, then the term *construction* has the apt connotation of a process of agents *building* from a number of different components. This is the sense explored in chapter 4 under the label "heterogeneous construction." A similar term, *heterogeneous engineering*, was introduced by Law (1987). Indeed, since the mid-1980s, the literature interpreting science has included many rich descriptions of the diversity of things scientists do and use in the process of making science: scientists employ or "mobilize" equipment, experimental protocols, citations, the support of colleagues, the reputations of laboratories, metaphors, rhetorical devices, publicity, funding, and so on (Latour 1987; Clarke and Fujimura 1992b, 4–5).

Nevertheless, in such descriptions—even the least colored ones—there are causes or "becauses" implied in the selection and juxtaposition of different factors. At this level of explanation, heterogeneous constructionism has not been well developed. Chapter 4 explores heterogeneous constructionism as an explicitly explanatory project, one that addresses the challenge of analyzing which of the diverse things mobilized make a difference and how they are combined to do so (see note 17).

Constructionists of various stripes (Sismondo 1993b; Hacking 1999) have offered explanations of scientific developments. Most of these explanations have, however, tended to move away from heterogeneity and from the implication that construction is a process (Taylor 1995b). Both these tendencies are especially evident when discussions associate social constructionism with society, context, ideology, and so on—something external to science—determining, penetrating, or being reflected in the content of accepted scientific theories. (The diagrams in Wise 1988 provide an explicit instance of this.) The resulting science then corresponds to the society in which it is generated or accepted. Conversely, when "realists" or anti-relativists dispute that the outcome of scientific activity—established knowledge—corresponds to society, they hold that it must then correspond—at least, generally or eventually—to nature or reality. The agency of scientists is not significant in these correspondence relationships. In the former, scientists can be seen as ciphers for society or dupes for interests; in the latter, they can be forgotten once they have helped to establish the knowledge.

Admittedly, published work is usually subtler than seminar or barroom discussions of social construction. The literature generally presents the society-science relationship as refracted and allows for the observation that not all of social group X believe Y and not all believers of Y come from social group X. In this vein we can see, for example, that scientists produce and judge knowledge mostly according to how it furthers goals of their social group (Shapin 1982). In short, agents are active and practically engaged. Nevertheless, construction in the simpler, correspondence sense has not been banished from such accounts. If we ask how these accounts explain why *this* knowledge was accepted and not *that*, and how *this* knowledge was generated in the first place, the implicit "because" seems, more often than not, to be one of correspondence between knowledge and interests (Woolgar 1981; Pickering 1993).

As chapter 4 conveys, the conceptual resources needed for heterogeneous constructionist explanations should be sought in thinking about process and practice (see note 15), not drawn from discussions of a reflection or correspondence relationship, however refracted or approximate that relationship is. Moreover, so as not to perpetuate the privileged status established knowledge has had in science studies, the separation of knowledge from scientific work and activity needs to be dissolved (Clarke 1991). Finally, the actions of agents who heterogeneously construct can never be governed solely or predominantly by the pursuit of any unitary goal, whether that goal is revealing the nature of some underlying reality (truth), establishing instrumentally reliable knowledge (Boyd 1991, 207; Hacking 1983), furthering the interests of the agent's social group, or maximizing and concentrating the agent's social resources (Latour 1988b, 160). The challenge—one shared with social theory—is to develop accounts of agency in terms of widely distributed causality (see theme 4 in chapter 4; chapters 5 and 6).

15. Practice

Attention to practice—to what scientists actually do—was a key development in the interpretation of science during the 1980s. This development is covered well by the collection of essays *Science as Practice and Culture* (Pickering 1992b) and in the editor's introduction (Pickering 1992a). Pickering adds the term *culture* to denote the "field of resources that practice operates in and on" (1992a, 2), and stresses the importance of considering the temporal aspect of practice—the process of making science as opposed to the products. Latour (1994) also explores a process metaphysics for interpretation of science, which follows the philosopher A. N. Whitehead. Even though the two cases in chapter 4 do not trace the course over time of the researchers' work, they share an emphasis on process and scientific practice. Heterogeneous construction might be read as Pickering's (1993) "mangle" and "impure dynamics," and the imagination of scientists as his "modeling" (see theme 3 on imagination in chapter 4; note 16). However, although Pickering shares many of the concepts outlined in the previous note, he theorizes practice mostly in terms of experimental practice (with one foray into conceptual practice). In an effort to distance himself from previous work that focused on sociological explanations of scientific knowledge, he diminishes the role of wider social resources and avoids discussion of causes. For Pickering, it is mostly because scientists tinker with tangible objects, whose resistance requires accommodation, that their goals and interests are subject to ongoing revision. Studies of practice in this sense are reviewed by Golinski (1990). In contrast, mathematical models or representations produced by researchers form the entry points in chapter 4; interpreting them motivates my analyses. I also remain interested in causes and explanation that span different domains of social practice (see also notes 14 and 17).

Lynch (1993) presents a strong challenge to the aspiration of explaining science. From an ethnomethodological perspective, he argues that sociological analysts cannot secure a vantage point that enables them to remain outside the vernacular language and epistemic commitments of the communities studied. How particular results stand with respect to prior results, how a laboratory's findings contribute to the disciplines, and so on are settled for all practical purposes by locally organized, embodied practices of handling equipment, making experiments work, presenting arguments in texts or demonstrations, and so on. When such practices are observed carefully—ethnomethodologically—they bear little relation to practitioners' accounts of what happened and why. By implication, interpreters of science can do no better.

My response to this challenge is to note that descriptions are not explanatorily innocent (note 17) and that ethnomethodological descriptions privilege explanations of the action of individual agents that assume that extra-local or trans-local considerations have no effect on their mental calculating or imagining (theme 3 in chapter 4; note 16). It is not clear that this lack of effect—especially on unconscious mental processes and embodied physical responses—could be demonstrated ethnomethodologically. Interrogating individuals about their motivations and memories and asking them to display them to audiences would not resolve the question because practitioners, as Lynch argues, are not privileged explainers of their own practice.

Turner (1994) strongly criticizes the project of interpreting science in terms of practice, but his objection is not to interpreters looking at what scientists do, but to their explaining knowledge by invoking hidden, but shared, premises or knowledge embodied in practices or routines. This objection does not seem pertinent to interpretations of knowledge-making that tease out multiple, practical considerations.

16. Imagination and the psychology of agents

Associating *imagination* and the *labor process* is Marx's idea. See *Capital*, vol. 1, pt. 3, chap. 7, sec. 1 (reprinted, e.g., in Tucker 1978, pp. 344–45). Robinson (1984) provides a relevant discussion of this passage.

The convention in social studies of science has been to avoid reference to an agent's psychology for fear of shifting the terms of explanation from the social realm to an unobservable realm of the agent's mind. I find dubious both the equation of social with observable and the empiricist rejection of unobservables. In any case, notice that imagination relies on a distributed, not an internal, notion of mind and psychology. Furthermore, psychological or cognitive models of the scientist as social agent are implicit in every explanation of the outcome of scientific activity. For example, Latour (1987) depicts scientists building "networks" in response to the stimulus of others building competing networks, and assumes that scientists seek to accumulate resources, all of which results, if successful, in "centers of calculation," "obligatory passage points" (Callon 1985), and their becoming macroactors (Callon and Latour 1981). Like the psychology of pigeons in the accounts of behaviorists, the psychology implied is both strong and minimal: the scientists are governed only by this egocentric metric of resource accumulation; they are not assumed to have multiple projects in their lives and work. This, like most other models of psychology and rationality implicit in social studies of science, is quite restrictive, even when rationalized as a methodological tactic intended to highlight the flexibility of agents' actions and network building (Taylor 1993).

17. Causes, explanation, and nonpartitionability

Cause and *explanation* are vexed terms in interpretations of science (Woolgar 1981; Latour 1988b) as they are in social science more generally (Lloyd 1986; Miller 1991). I am interested enough in the sense of causality and explanation reflected in chapter 4 to explore it throughout the rest of the book, but I also recognize the importance of getting other interpreters of science interested in paying attention to scientists' diverse resources (see chapter 5). This note, therefore, is intended to help readers position my account in relation to their own and other positions. It is a lengthy note, but to have included it in the body of the text would have significantly interrupted the main path of development. On the other hand, to shorten or omit this discussion would have reduced the chance of my attracting some sympathetic philosophers to relate my position to the terms of established philosophical ones (e.g., Mackie's [1965] INUS conditions, Harré and Madden's [1975] causal powers). Yet this note avoids technical philosophical terminology, aiming to meet the needs of some general readers as well as those of philosophical specialists. (One specialist reference with particular affinity to the formulations to follow is Hart and Honoré 1959; see, in particular, the introduction and first two chapters.)

Descriptions favor certain explanations over others; by selection and juxtaposition, they give weight to different factors and imply that some things happen *because*

of previous and ongoing things. And if there is a "because," then a how or why question is being answered—that is, an explanation is being given. This said, I need to
distinguish my formulation of causes and explanation from many others. I am not
referring to the big-C Causality underlying grand trajectories of social development—
as in modernization theory or versions of Marxism—or to causes as the source of
deviations in specific societies from those essential trajectories (McLaughlin 1989). My
model of causes is not the physical sciences' fundamental causes exposed one by one
through suitably designed, repeatable, controlled experiments; nor do I want to
promote explanation in the sense of statistical regularities. (The cases in chapter 4
should indicate my willingness to explain singular, nonrepeatable situations; see Lloyd
1986 and Miller 1991.) Finally, I am not interested in covering law generalization-
abstractions for interpreting science. This last formulation is the focus of Latour's
(1988b) polemic against explanation in general, in which he opposes tying a range of
outcomes in one variable realm (science) to some feature of a relatively stable realm
(society). Latour does not see social life as stable or as a realm separate from science.
He wants to highlight the novel coalitions and outcomes involved in the production of
science and society. Although I agree to a large extent with this perspective, the very
network accounts he advocates build on multiple, diverse causes (Latour 1987).

Causal analysis of heterogeneous webs, as I formulate them, proceeds in the spirit
of historical explanation. That is, the analysis must attempt to identify numerous
causes, no one overshadowing the others, but each in context making a difference
and all together providing a composite or conjunction of conditions sufficient for
readers to see why or how the situation or outcome under consideration, and not
some other possibilities, happened (Miller 1991; Taylor 1994a). In this sense, explaining is synonymous with giving an account of causes—that is, of conditions that
would, if changed, make a difference to the outcome.

Some elaboration and refinement are called for. Causes and counterfactuals are
bound together (see also note 20); explainers choose to address certain contrasting
possibilities and not others, and to allow for certain conditions to be changed and
other conditions to be backgrounded because they are fixed or taken for granted.
An audience specificity tends to be imported by such explanatory choices. That is,
although the hypothetical "if changed" need not be construed as the condition actually being changeable, acceptance of an explanation is enhanced, in practice, if an
audience can be found or enlisted that considers that the condition is (or was)
changeable. (Causes and realizable changes, not just hypothetical counterfactuals,
tend to be bound together.) Explanations in the sense of sufficient composites should,
in light of explanatory choices and their audience specificity, always be viewed as
provisional. They are subject to competition from other composites and likely to be
superseded if the categories and detail of one of the others allows the reader to imagine more intimately how the agents were acting in the given situation (Foucault 1981;
Humphries 1990).

Nonpartitionability of causes is an important feature of heterogeneous constructionist explanations (see item 8 on the list at the end of chapter 4). Given a number of
focal causes, each of which makes a difference in context—that is, in conjunction with
other causes and background conditions—any analysis of the effect of a cause must

operate jointly, not cause by cause. But, taken to the extreme of "everything is contingent on everything else," this joint causation would be impossible for explainers and their audiences to handle. In practice, explainers discount some conditions as incidental, focusing, for example, on the modeler's choice of software but not, say, on his use of baking soda for toothpaste. A variety of factors can take simplification further. For example, explainers can combine focal causes or background conditions into synthetic or structural conditions (e.g., centralized government policymaking); they can minimize the number of focal causes by shifting some or most to the background or incidental categories; they can focus on causes of a similar kind (e.g., ones relating to farmers' decisions and not ones that span the international economy and politics). This homogenization can even go so far as to lead to gross dichotomous categories of causes, such as natural and social, which are then held to "interact"—a bland term, often a smoke screen for the belief that one category is important, the other incidental. Heterogeneous constructionism, however, does not want to homogenize to that extent. While simplification is necessary for explanation, there is no logical or empirical reason for not choosing an intermediate level of simplicity or complexity, in which the causes remain multiple and of diverse kinds. (The idea of diverse kinds assumes that we still permit ourselves the linguistic convenience of classifying into distinct kinds causes that are intricately interlinked.) At this level, it is difficult to "forget" that causes are causes-linked-in-context, and thus it is difficult to partition relative importance or responsibility for an outcome among separate types of causes. The character and implications of intermediate complexity (chapter 5, section C), in which explanations preserve heterogeneity of causes and their interlinkages, warrant more attention from philosophers.

Finally, note that nonpartitionability does not mean that all causes are equally important. Instead, it means that causes are linked in context, so that the size of a cause's effect is conditional on the given interlinkages. Similarly, something is not a resource until it is linked or mobilized. In principle, this conditionality includes backgrounded conditions, but in practice, as long as change in the background conditions is not being considered, the conditionality tends to be omitted from accounts.

18. "Natural" ecology versus socio-environmental research

My involvement in ecology originated in environmental activism (see the prologue), so I have not been troubled by moving between the study of "natural" ecology—ecological situations that seem relatively undisturbed by human societies and do not include humans as a component—and socio-environmental situations. Indeed, as I was working on this book, many other ecologists began to accept that this boundary was difficult to maintain (e.g., Holling 1995). Environmental historians observed that the nature of environmental change has always been an outcome of social interactions and negotiations (e.g., Cronon 1983). Together with interpreters of science (e.g., Yearley 1991) and political ecologists (e.g., Peet and Watts 1996b), they also drew attention to the social interactions and negotiations that shape how knowledge about society, ecology, and socio-environmental relations is made.

19. Adaptive Environmental Management and policy experimentation

As stated in the text, Adaptive Environmental Management (AEM) begins with a recognition that the dynamics of any ecological situation are not fully captured by any model or composite of models, especially because management practices produce continuing changes in those dynamics and make the ecological situation a moving target. AEM turns that limitation into an opportunity, attempting to bridge gaps in knowledge through carefully designed experiments in environmental management. In these experiments, a range of management practices, chosen on the basis of existing knowledge and model-based predictions, are implemented. Lessons about the practices and models are drawn from the different outcomes (Holling 1978; Walters 1986; Lee 1993; Gunderson, Holling, and Light 1995a; Ebata 1997). The focus of AEM is on the dynamics of species other than humans, such as fish that spawn in tributaries of dammed rivers, but social institutions are, of course, involved in the implementation of alternative management practices—or in thwarting or confounding such experiments (Lee 1993; Walters 1997). AEM has begun, therefore, to include models of the complexity of the social situations in which the experiments are implemented (Berkes, Colding, and Folke 2002).

AEM researchers are often active participants in these social situations, but their socio-environmental models have often been system-like. These models seem to position the researchers as outside observers seeking "to reveal the simple causation that often underlies the complexity of time and space behavior of complex systems" (Holling 1995, 13; see also Gunderson, Holling, and Light 1995b). Researchers and others who understand the dynamic relationship between institutional behavior and ecological degradation have a special role in breaking "decision gridlock" and bringing about "renewal of ecosystems and institutions" (see The Resilience Alliance, n.d.). (This might be viewed as another example of a philosophy framed in terms of universal interests that builds a special place for its proponents in the proposed social organization; see chapter 3; chapter 6, section B2.) In practice, of course, many kinds of agents—scientists, managers, and other stakeholders—have to be involved in formulating and conducting appropriate experiments. More recently, some work in AEM has included "participatory approaches to stakeholder-driven analysis" (The Resilience Alliance, n.d.; see also Lee 1993).

Chapter 5

20. Counterfactuals: Typical objections and some rejoinders

Conventionally, counterfactual analysis has not been endorsed by historians and social scientists. This note addresses some of their typical objections. The first is that, given the multiplicity of components present in any web—a nearly infinite number of alternatives exist—how does one choose the relevant subset? For example, in chapter 4, section A, why did I stop at multi-objective techniques? Why not analyze farmers' decision making in terms of, say, neural nets or genetic algorithms? My answer is that, when analysts use counterfactuals to expose resources and to support the corresponding causal claims, they must make the counterfactuals *practically* plausible. The set of counterfactuals should not include just any conceivable idea. To persuade readers that a counterfactual was practically plausible to the agents, the counterfactual should not be too dissimilar from what actually occurred. Or, more precisely, the counterfactual should, given the cross-linkage of resources, build on most of the same resources as the actual situation.

Now, evaluations of whether resources linked into the altered (counterfactual) context are still the same cannot be neutral, but must be made by, or with reference to, two groups: some agents to whom the alternatives are also practically relevant, and some audience that has to imagine how the agents were acting in the given situation. Using such an evaluation, a finite subset of alternatives can, in principle, be delimited in a nonarbitrary way. Hawthorn (1991) provides three detailed cases to illustrate his valuable discussion of the use of counterfactuals in historical explanation.

The need to show practical plausibility of alternatives can lead to a second objection: To do this well, one should have a strong picture of what the relevant causes are, yet the very reason for counterfactual analysis is to assess the causal significance of resources. This apparent circularity dissolves, however, if we think of explanation as an iterative process, beginning with causal ideas (see note 17) borrowed from situations deemed similar. These ideas are then successively refined and reformulated to address the situation at hand. It is true that iterative methods cannot in general guarantee that one successively approaches a correct account, but such certainty need not be a decisive criterion for good interpretation.

Lynch (1989), following Elster (1978), advocates a different way of resolving the circularity problem; namely, by building a more explicit theoretical basis for counterfactuals from other sources, such as the sociology of scientific knowledge. Besides that of Hawthorn (1991), who eschews generalizing theory, the most developed use of counterfactual scenarios has been in economic history, in which explicit econometric models have been used to examine issues such as the effect of railroads on the U.S. economy in the nineteenth century (Fogel and Engerman 1969). This work takes seriously the idea of going back in time until there is a point at which counterfactuals (e.g., greater development of canals as a counterfactual to railways) can be smoothly inserted into the model (see below). Notwithstanding this virtue, econometric counterfactual analysis does not provide satisfactory exemplars of theory-based counterfactuals. Because econometric scenarios are formulated as sets of regression equations, their "causality" is based on statistical association and is difficult to express in terms of actions of agents. On neither count does this match the view of causality here (see note 17).

The specific example of the KFM in chapter 4, section A, presents an easy case for convincing readers that the counterfactuals were practically plausible and relevant. That is because the modeler actually attempted to pursue or envisage the alternatives that were mentioned. Neural nets and genetic algorithms, on the other hand, were in their very early days of development, in disciplines and places far removed from the economics of the Institute. However, we should not make too much of the special insight the modeler provided. Counterfactual analysis must be possible regardless of whether the agents have attempted to implement alternatives. For this reason, the Kerang study and the modeler's engagement should not be taken as the exemplar of how to expose and identify alternatives. Observations that were more sociologically distanced or systematic than those the modeler provided would help make the analysis something others could borrow from. To make systematic choices among the multiplicity of counterfactuals, it helps to be prepared to expose and develop one's own agenda for engagement. In formulating the set of eight contrasts in chapter 4, section B, my interest in participatory rather than technocratic approaches to socio-environmental studies came into play. This interest made it clear why I discounted Cockrum's alternative to Picardi's system dynamics modeling and instead examined alternatives in which the pastoralists' situation became less system-like.

Finally, some observations can be made on why interpreters of science might want to depart from the naturalistic approach of staying close to the scientist's vantage point. As mentioned earlier, counterfactual analyses are strongly shaped by the positioning of the sociological explainers and their audience. A counterfactual that the audience at hand does not consider practically plausible for the relevant agents in its time can sometimes be suitably modified. By moving back in time, to another place, or among different people, we can render it into something practically plausible for the new agents. (But we may also need to move our audience, or move in front of a different audience, in order to do so.) To help explain the Kerang study, we could ask whether a different modeler—say, one more senior and more experienced in multi-objective techniques—could have made a different model "right for the job." This would mean going back to the time when the project was formulated, or even further, to the time when other scientists were devising the courses on modeling that went into the training of this different modeler. A shift in the counterfactuals and their point of insertion—the time, place, focal agents, and audience—would also be required if we wanted to test a hypothesis, say, that policies formed through studies conducted in-house by a government agency are more likely to be taken up than policies formed through research contracted out to universities. Clearly, it is a challenging task to define and delimit counterfactuals when agents are building on webs of heterogeneous resources. This task is a facet of the broader challenge of analyzing the composition, shape, and structure of those webs.

21. Reflexivity

To be reflexive is to apply one's analytic approach to one's own work. The call for interpretive reflexivity became a popular theme in sociology of science during the 1980s (Woolgar 1988; Ashmore 1989). This theme followed from, but filled out considerably, the conventional wisdom that scientific work is more readily accepted if related to the current preoccupations and frameworks of one's discipline (Kuhn 1977). The emphasis in reflexivity studies has been on the textual and rhetorical strategies used to advance an argument or interpretation. Reflexivity can, however, be extended to (1) the practical considerations involved in establishing any argument or interpretation—the focus of chapter 5—and (2) the kinds of engagements that interpreters of contemporary science have in the controversies they study. This second area has been the subject of especially active debate (Scott, Richards, and Martin 1990; Ashmore and Richards 1996). Some scholars on the side of open partisanship, such as Martin (1996a, 1996b) and Hess (1997, 150–52), advocate deliberate attempts to perturb the dynamics of scientific endeavors in the course of investigating those dynamics. Yet, from the perspective of heterogeneous construction, even when interpreters of science are not overtly partisan, they are contributing to different social actions. This makes them responsible—albeit partially and jointly—for those actions. Strong grounds for practical, not just textual or rhetorical, reflexivity are provided by combining this responsibility with a desire for consistency between how one views the agency of scientists and one's own agency and with curiosity about the construction of one's own work.

22. Mapping workshops and teaching

One participant in the Helsinki mapping workshop observed that "one question leads to ten questions—What is a lake? Why is phytoplankton one category? . . . " Another likened mapping to the process of writing and revising: "Like writing, out of mapping comes awareness of new parts of the map that need more work." Just as in a good graduate student dissertation research seminar, students raised a whole range of issues—from nagging uneasiness they feel about certain research directions to specific technical points—that would have remained latent in a seminar dedicated to a specific theme. Mapping workshops certainly warrant attention from other teachers as an approach to stimulating advanced students to define their research (Taylor and Haila 1989). This approach, it should be noted, differs markedly from concept mapping (Novak 1990), in which the focus is on well-established relationships between concepts. Mapping as described in chapter 5 allows mapmakers to explore what is not yet clear about an issue on which they want to take action.

23. Process

Ideas about discontinuities and transitions often rely on a sense of the term *process* that I am avoiding; namely, a basic underlying causal structure that allows people to explain events as instances of the essential process or as noisy deviations from it. I use the term *process* in the sense of sequences of events that persist or are repeated sufficiently long for us to notice them and need to explain them. Such structuredness, in combination with differentiation and historical contingency, distinguishes my account (and political ecology more generally) from Vayda and Walters's (1999) more particularist and skeptical-of-theory approach, with which it otherwise shares many qualities. Partially in recognition of essentialist connotations of process, I have sometimes used the neologism "*intra*secting processes" to convey the inseparability of the different strands.

Chapter 6

24. A tragedy of the commons class simulation, part 1

For simplicity of making calculations while running the simulation, I set the threshold for overgrazing at 100 cows. Below this threshold, I make the income per cow per year $100, and above it, $200 minus total herd size. I do not inform herders of the threshold or the formula I use to calculate the income; they make their own sense of the trend. If the class has N members, I set the initial number of cows per person at about $80/N$. I use the same figure as the maximum number of cows a herder can purchase in any one yearly cycle and, multiplied by $100, set this as the initial cash per person. That is, if $N = 20$, herders begin with 4 cows and $400 each. The buying price per cow I set at $100. Herders indicate purchases on pieces of paper. I add these up as I collect them and then update the total herd size and income per cow on the board. Herders then update their accounts and decide on the next year's purchases. I am also a herder, and, although I do not tell anyone, I purchase as many cattle each year as allowed, which ensures that overgrazing will occur. When we review the situation, I ask all the students to mark their herd size and monetary wealth on a graph on the board so that the disparities among the herders are apparent.

This simulation shares some features with other tragedy of the commons simulations (e.g., Holle and Knell 1996; Mitchell 1997). The goal in my class, however, is not to reinforce the idea that the tragedy is real and important for understanding environmental degradation.

25. A tragedy of the commons class simulation, part 2

The standard post-Hardin lesson is that agents communicating and working together in communities can overcome their short-term self-interest and build local institutions for managing a resource held in common. Successful institutions are operated by those directly concerned with the resource and are "externally accepted"; that is, the government, markets, or industries tolerate, or even support, the community of users' jurisdiction over the resource (Berkes et al. 1989; Ostrom 1990). In the future, I plan to extend the simulation in the direction of these post-Hardin lessons. My idea is to proceed until the situation of overstocking and unequal assets develops, but instead of continuing negotiations as a whole class, I would divide the class into groups of three, each with one person well-off, another poor, and the third in-between. Each triad would be given a particular scenario (e.g., nomadic pastoralism, Western U.S. cattle grazing, New England fishing) and be asked to try to negotiate a mutually acceptable arrangement among the three persons. After a while, I would ask them to respond to a little devil who whispered in their ear: "Add another cow to your herd or pull in another fish—you'll get all the benefit and any cost will be shared by all." (I suspect that the responses will be more qualified and contingent than Hardin implies.) The simulation would end with reports back to the whole class on (1) the group's negotiations and outcomes and (2) different individuals' responses to the little devil.

26. A tragedy of the commons class simulation, part 3

In a variant of the class simulation, I pose a pair of questions to elicit the actions or policies that correspond to the different formulations and the kinds of agents involved. For each formulation, I ask the students: What is to be done on the basis of this science? What more would you like to know than this science shows? This exercise correlates the different formulations with broad orientations to policy (see Taylor 2001d). Moreover, policies that fail to have their desired effects indicate a need for a different formulation of the science. During the 1970s and early 1980s, for example, views about overstocking of the range led to development projects whose goal was to produce fundamental changes in pastoral practices—privatizing pasture, reducing stocking rates, and moving pastoralists into large-scale ranching schemes. These projects generally failed; research led belatedly to the perspective that herders respond skillfully and sensitively to their variable and uncertain semiarid environments, provided that they can remain mobile, maintain species diversity in their herds, and apply their local ethno-sciences of range management (McCabe and Ellis 1987; Horowitz and Little 1987).

27. Theory about social structure and agency in the context of environmental studies and interpretations of science

Correlations between knowledge and social interests. See introduction to chapter 3, section A; notes 11, 12, and 14 and references therein, especially Woolgar's (1981) critique of interests explanations. Taylor (1993) interprets the aversion to social explanations of science evident in Woolgar's (1981) and later accounts in terms of implicit, interconnected views about the psychology of agents, social causality, the structuredness of society, and the role of agents in the reproduction of social structuredness (see *Agency and structuredness* below).

Social theory. Traditional, big "S" Social theory seeks to account for the structure and dynamics of Society as a whole (Münch 1987; see also edited collections Bottomore and Nisbet 1978; Giddens and Turner 1987). Although such theory is a possible source of propositions to inform researchers' accounts of their situatedness in society, modern Social theory itself provides grounds for a critique of its own project. Illuminating this point, Goldblatt (1996) examines the contributions that Social theorists Giddens, Gorz, Habermas, and Beck make to shaping plausible, politically appealing, and practical institutional alternatives and innovations in the context of environmental degradation and the rise of environmental concerns in Western politics. Among the many respectful criticisms Goldblatt makes of these theorists' work is his observation that the globalization of capitalism and (following Giddens and Beck) reflexive modernization mean that "[t]oo many decisions about economic rationality have to be made by reflexive agents on the ground, on the basis of tacit practical knowledge, to make the transfer of decision making powers to the centre effective. No state, however flexible, can gather enough information, process it quickly enough or embody the essentially local knowledge and skills required in a rapidly changing economy" (Goldblatt 1996, 193). It follows, I believe, that no theory about the dynamics of Society as a whole could provide sufficient resources for reflexive researchers. Researchers may find it helpful to consider multiple, partial social theories, but the challenge remains of weaving those theories together so that researchers do not allow simple propositions about overarching or underlying processes (see note 23) to govern their accounts of social situatedness (Taylor 1997a, 211ff). (For other accounts of social theorizing in the context of environmental change, see Harvey 1993; Peet and Watts 1996b; Redclift and Benton 1994.)

Agency and structuredness. There has been a long history in social theory of discussion of how to relate social structure and human agency (Dawe 1978; Giddens 1981; Sewell 1992; Vogt 1960; see Taylor 1996b for bibliography in the context of interpretation of science). Concepts introduced in *Unruly Complexity* provide the basis of a framework for moving beyond the structure-agency dualism. In brief: Envisage agents operating within intersecting processes (IPs) that are interlinked in the production of any outcome and in their own ongoing transformation (chapter 5, section C). Let these IPs be teased out into three sets of three IPs: the *Personal,* which connects the IPs of cogitation, body, and unconscious; the *Local,* which connects discursive themes, materials at hand, and local rules; and the *Social,* which connects Discourse, Materiality, and Rules. Agents heterogeneously construct a variety of projects at any time. In doing so, they imaginatively mobilize discursive themes, materials at hand, and local rules. Their cogitation involves some thematic framework that simplifies their actual and possible heterogeneous construction as it is constrained and facilitated by their unconscious and body. The Local IPs evolve as an outcome of what different agents are able to do in response to what other agents are doing. The Social IPs evolve as the linkage of many Local IPs and are, in turn, drawn on or invoked through discursive themes by interacting agents in Local IPs (see reflection on vibrating agency in the text above).

Such a framework makes conceptual room for a view of distributed agency in relation to social *structuredness*. There is no reduction to macro- or structural determination. Nor is the focus on transactions among concentrated individual agents. Even if agents tried to stay focused on following some principle of morality or rationality, or sought to optimize some metric, such as their profit, they could not avoid contributing to many projects, given the intersections among Personal, Local, and Social IPs. The view of human nature implied by the framework is similar to that of Dervin (1999), in which agents try to bridge "gaps" opened up by the inherent incompleteness or unboundedness of reality and by their movement in time-space. Contingency is unavoidable, even necessary, in psychological development and construction (Hendriks-Jansen 1996; Urwin 1984).

The framework also resists the subordination of the material to the mental or discursive that is effected, for example, by sociologist of knowledge Barnes when he equates social order with "shared knowledge and aligned understandings" that confer "a generalized capacity for action upon those individuals who carry and constitute it" (Barnes 1988, 32, 57), or by social epistemologist Fuller when he analyzes the rhetoric of promoting "public understanding of science" and calls for experimentation in widening public participation in debates over scientific claims (e.g., Fuller 2000). (Markus 1986 provides a general analysis of the difficulties that philosophers and social theorists have in reconciling the paradigms of "language and production.") Of course, my framework specifies nothing about the particulars of any situation or how different agents should engage within those particularities, leaving most of the work still to be done.

In social studies of science it has become popular to invoke nonhuman agency, a move initiated by Latour and Callon when they used the semiotic label "actants" for humans, other living beings, and nonliving things alike in their descriptions of how scientists secure support for their theories (Callon 1985; Latour 1988a, 1999, 2004). The playfulness of the resulting anthropomorphic accounts seems to animate the discussion of the nonhuman resources, but in practice, Latour's and Callon's accounts reduce agency to a lowest common denominator; namely, resistance to the agency of others. Human purposes, motivations, imagination, and action do not enter the analysis, except that humans have to attempt to overcome resistance. Taylor (1993) interprets this move as follows: If scientific agents are viewed as acting with a minimal psychology—almost without mental representations—then this ensures that inborn dispositions, cognitive constraints, individual creativity, and so on cannot determine action and belief. This absence preempts the analyses of others who invoke the internal cognitive mind to resist the social construction of science. It also leaves no place for interests or other external influences to reside inside the scientist's head, and thus counters earlier analyses in social studies of science that allowed social context or social forces to determine scientists' beliefs or actions. In short, invoking nonhuman agency can be interpreted as promoting a particular view about social causality and the character of human agency in the production and reproduction of social structuredness. (See Downey and Dumit 1997 for alternative perspectives on nonhuman agency, which begin from observing anthropologically the routine practices in which people—not only interpreters in social studies of science—treat technologies and other things as agents.)

28. **Gender and race in relation to social structuredness and interpretation of science**

As observed in note 4, almost all the researchers in my cases are men of European descent, which invites stretching (or critique) of my work in relation to the gendered or racialized dimensions of science. At the same time, heterogeneous constructionism may stretch (or complement) feminist and anti-racist interpretations of science and technology. To explicate this claim, first let me categorize into three levels the gender dimensions of social structuredness in relation to science and technology. (Equivalent levels can be articulated for differences that refer to race, ethnicity, or European descent versus other othernesses.)

1. *Underrepresentation of women* in science and in technological design; obstacles to and underrecognition of their contributions; possibilities for women's standpoint to address aspects of the world underrecognized by men.
2. *Biases* in knowledge and technologies that claim to represent progress, efficiency, or other universal interests, but in practice promote the unequal social status of men over women.
3. The pervasiveness of *gender-like dualisms* in which one category is subordinate to the other and complex spectra are purified into dichotomies; the suppression of ways in which these conceptual schemes are troubled by multiplicities and hybrids.

Now, from the perspective of heterogeneous constructionism, *some* of the resources that are mobilized in establishing specific knowledge or technologies will be drawn from these different levels of structuredness. But it is implausible that *all* the resources mobilized would derive from gendered social structuredness, let alone from only one of the three levels. The challenge becomes to analyze the ways in which diverse resources are linked over time by knowledge-makers in particular constructions and to act on the basis of such analyses (chapter 6, section C). A very significant source of resources has been the existence of a feminist movement(s) based on a broader set of social and personal concerns, which continues to bring attention to issues about science and technology on all three levels (Keller 2001; Schiebinger 2003).

29. **Practices of reflection and group interaction**

The principle that "we know more than we are, at first, prepared to acknowledge" (Taylor 2002b) not only emerges from mapping workshops, but also lies behind my use of opening-up themes (type 2 formulations) to strengthen the vibrations in the direction of particular dynamics (type 3 formulations). It is a principle also shared by several practices of reflection that can fruitfully be injected into teaching and research workshops (Taylor 2001b; Taylor 1997d compares several related schemas for learning and group process):

- *Focused conversations* and other participatory workshop processes developed by the Institute of Cultural Affairs bring insights to the surface and ensure that the full range of participants are invested in collaborating to bring the resulting plans or actions to fruition (Stanfield 1997; Taylor 1999a; see the epilogue, section B).
- *Freewriting* allows participants to acknowledge other preoccupations first and thus clear mental space so that thoughts about the issue in question that had been below the surface can emerge (Elbow 1981; Taylor 2002b).
- *Constructivist listening*, based on reevaluation counseling, allows participants in pairs to delve deeper into emotions left from hurtful experiences that interfere with clear thinking, making sense of experience, and listening well to other participants (Weissglass 1990).
- The *Sense-Making* approach to identifying and bridging gaps between what people know and need to know also helps people to reevaluate their customary or habitual responses, to acknowledge the emotional valences in scholarly and day-to-day interactions without becoming blocked by the emotions (Dervin 1999). (This approach corresponds to a theory of human agency with strong affinities to that developed in this book; see note 27.) Research on Sense-Making indicates—among other findings—that people make much better sense of scholarly presentations when they are contextualized along the following lines:
  a. The essence of the project is . . .
  b. The reason(s) I took this road is (are) . . .
  c. The best of what I have achieved is . . .
  d. What has been particularly helpful to me in this project has been . . .
  e. What has hindered me has been . . .
  f. What I am struggling with is . . .
  g. What would help me now is . . .
- The *dialogue process,* in which participants speak and listen without disputing or judging the diverse points of view expressed by others, allows meaning to evolve collectively as participants' thought processes become more visible or accessible to one another and to themselves (Isaacs 1999).

Epilogue

30. Narrative that keeps distributed agency in view

Kondo (1990) exposes the challenges she faced as an anthropologist studying and writing about de-centered agency. Therapists White and Epston (1990) encourage their clients to articulate previously subordinate narratives so as to move beyond the family and social dynamics that perpetuate dominant, constraining narratives. The new narratives are "not radically invented inside our heads. Rather, [they are] negotiated and distributed within various communities of persons and in the institutions of our culture" (White 1998, 225). The implications for physical as well as psychological distress of articulating new narratives are discussed in Griffith and Griffith (1994).

31. **Participatory action research and collaboration in environmental management**

In participatory action research (PAR), social scientists shape their inquiries through ongoing work with and empowerment of the people most affected by some aspect of economic or social change. Greenwood and Levin (1998, 173–85) review various approaches to PAR (see also Park et al. 1993; Selener 1993), but the power of participation is better conveyed by accounts of the struggles of local peoples to influence science and politics (e.g., Adams 1975; Gibbs 1982; Brown and Mikkelsen 1990). A current of PAR, albeit a minor one, can be found in the interpretation of science and technology (Martin and Richards 1994; Taylor 2004). PAR and its cognates are most widely promoted in rural development in poor areas of the world, from which cases are often drawn to illustrate the rise of citizen participation and of new institutions of *civil society* (Burbidge 1997).

Participation is not always invoked with the sincerity or the success evident in the CARE project described in the epilogue. Indeed, the mandate for participation can be wielded in disempowering ways by state or international agencies (Agrawal 2001; Ribot 1999; see also Peters 1996 for a review of the politics of participation and participation rhetoric). Nevertheless, in industrialized countries as well as in poor rural regions, environmental planning and management increasingly builds in stakeholder collaboration; that is, explicit procedures for participation of representatives of community groups, government agencies, corporations, and private property owners (Wondolleck and Yaffee 2000; Borchers and Maser 2005; see also note 19 and the epilogue, section B).

References

Abrams, P. A., and D. A. Taber. 1982. Complexity, stability, and functional response. *American Naturalist* 119:240–49.

Adams, F., with M. Horton. 1975. *Unearthing Seeds of Fire: The Idea of Highlander*. Winston-Salem, NC: John F. Blair.

Agrawal, A. 2001. State formation in community spaces?: The Forest Councils of Kumaon. *Journal of Asian Studies* 60 (1): 1–32.

Agrawal, A., and C. Gibson. 1999. Enchantment and disenchantment: The role of community in natural resource conservation. *World Development* 27 (4): 629–49.

Akin, W. E. 1977. *Technocracy and the American Dream*. Berkeley: University of California Press.

Allee, W. C., A. E. Emerson, O. Park, T. Park, and K. P. Schmidt. 1949. *Principles of Animal Ecology*. Philadelphia: Saunders.

Allen, T. F. H., R. V. O'Neill, and T. W. Hoekstra. 1984. Interlevel relations in ecological research and management: Some working principles from hierarchy theory. United States Department of Agriculture (Report RM-110): 1–11.

Ashmore, M. 1989. *The Reflexive Thesis: Wrighting Sociology of Scientific Knowledge*. Chicago: University of Chicago Press.

Ashmore, M., and E. Richards. 1996. The politics of SSK: Neutrality, commitment and beyond. *Social Studies of Science* 26 (2): 219–468.

Austin, M. P. 1980. Searching for a model for use in vegetation analysis. *Vegetatio* 42:11–21.

———. 1987. Models for the analysis of species' response to environmental gradients. *Vegetatio* 69:35–45.

————. 1999. The potential contribution of vegetation ecology to biodiversity research. *Ecography* 22:465–84.

Baldwin, E. 1947. *Dynamic Aspects of Biochemistry*. Cambridge: Cambridge University Press.

Barnes, B. 1988. *The Nature of Power*. Urbana: University of Illinois Press.

Bateson, G. 1946. Physical thinking and social problems. *Science* 103 (2686): 717–18.

Bechtel, W., and R. Richardson. 1993. *Discovering Complexity: Decomposition and Localization as Strategies in Scientific Research*. Princeton, NJ: Princeton University Press.

Bender, E. A., T. J. Case, and M. E. Gilpin. 1984. Perturbation experiments in community ecology: Theory and practice. *Ecology* 65:1–13.

Benjamin, A. E., G. N. Cantor, and J. R. R. Christie, eds. 1987. *The Figural and the Literal: Problems of Language in the History of Science and Philosophy, 1630–1800*. Manchester: Manchester University Press.

Berkes, F., J. Colding, and C. Folke, eds. 2002. *Navigating Social-Ecological Systems: Building Resilience for Complexity and Change*. Cambridge: Cambridge University Press.

Berkes, F., D. Feeny, B. McCay, and J. Acheson. 1989. The benefits of the commons. *Nature* 340:91–93.

Bhaduri, A. 1983. *The Economic Structure of Backward Agriculture*. London: Academic Press.

Black, M. 1962. *Models and Metaphors: Studies in Language and Philosophy*. Ithaca, NY: Cornell University Press.

Bloomfield, B. P. 1986. *Modelling the World: The Social Constructions of Systems Analysts*. Oxford: Blackwell.

Blum, A. S. 1993. *Picturing Nature: American Nineteenth-Century Zoological Illustration*. Princeton, NJ: Princeton University Press.

Boal, I. 2005. *The Long Theft: Episodes in the History of Enclosures*. San Francisco: City Lights Books.

Bocking, S. 1997. *Ecologists and Environmental Politics: A History of Contemporary Ecology*. New Haven, CT: Yale University Press.

Bohm, D. 1995 [1980]. *Wholeness and the Implicate Order*. London: Routledge.

Bookchin, M. [pseudonym: L. Herber]. 1962. *Our Synthetic Environment*. New York: Alfred A. Knopf.

Borchers, J., and C. Maser. 2005. *Understanding Constraints in Sustainable Development*. Chelsea, MI: Lewis. Forthcoming.

Botkin, D. 1990. *Discordant Harmonies: A New Ecology for the Twenty-first Century*. New York: Oxford University Press.

Bottomore, T., and R. Nisbet, eds. 1978. *A History of Sociological Analysis*. New York: Basic Books.

Boyd, R. 1991. On the current status of scientific realism. In R. Boyd, P. Gasper, and J. D. Trout, eds., *The Philosophy of Science*, 195–222. Cambridge, MA: MIT Press.

Brazil, W. 1975. Howard W. Odum, the building years 1884–1930. Ph.D. dissertation, Harvard University, Cambridge, MA.

Broad, W. J. 1994. Plan to carve up ocean floor near fruition. *New York Times* (March 29): C1.

Brokensha, D. W., M. M. Horowitz, and T. Scudder. 1977. *The Anthropology of Rural Development in the Sahel: Proposals for Research*. Binghamton, NY: Institute for Development Anthropology.

Brown, D. S. 1973. U.S.A. statement. In *Final report on the meeting of the Sudano-Sahelian mid- and long-term programme,* 93a. New York: United Nations, Special Sahelian Office.

Brown, L. R., C. Flavin, and H. Kane. 1996. *Vital Signs: The Trends that Are Shaping Our Future*. New York: W. W. Norton.

Brown, M. 1977. War, peace, and the computer: Simulation of disordering and ordering energies in South Vietnam. In C. Hall and J. Day, eds., *Ecosystem Modeling in Theory and Practice: An Introduction with Case Histories,* 393–417. New York: Wiley.

Brown, M. A. 2001. IB95010: The Law of the Sea and U.S. policy. www.cnie.org/nle/mar-16.html (viewed 11/15/01).

Brown, P., and E. J. Mikkelsen. 1990. *No Safe Place: Toxic Waste, Leukemia, and Community Action*. Berkeley: University of California Press.

Bryson, R. A. 1973. Drought in Sahelia: Who or what is to blame? *Ecologist* 3:336–71.

Bugher, J. C. 1970. Project Foreword. In H. T. Odum and R. F. Pigeon, eds., *A Tropical Rain Forest: A Study of Irradiation and Ecology at El Verde, Puerto Rico*. Oak Ridge, TN: United States Atomic Energy Commission.

Burbidge, J., ed. 1997. *Beyond Prince and Merchant: Citizen Participation and the Rise of Civil Society*. New York: Pact Publications.

Bush, V. 1945. *Science: The Endless Frontier*. Washington, DC: United States Government Printing Office.

Callon, M. 1985. Some elements of a sociology of translation: Domestication of the scallops and the fishermen of St. Brieuc Bay. In J. Law, eds., *Power, Action, Belief: A New Sociology of Knowledge?* 196–233. London: Routledge & Kegan Paul.

Callon, M., and B. Latour. 1981. Unscrewing the big Leviathan: How actors macro-structure reality and how sociologists help them to do so. In K. Knorr-Cetina and A. V. Cicourel, eds., *Advances in Social Theory and Methodology: Toward an Integration of Micro- and Macro-sociologies,* 277–303. Boston: Routledge & Kegan Paul.

Cannon, W. B. 1933. Biocracy: Does the human body contain the secret of economic stabilization. *Technology Review* 35 (6): 203, 204, 206, 227.

Carson, R. 1962. *Silent Spring*. Boston: Houghton Mifflin.

Caswell, H. 1988. Theory and models in ecology: A different perspective. *Bulletin of the Ecological Society of America* 69:102–9.

Chase, A. 1977. *The Legacy of Malthus: The Social Costs of the New Scientific Racism*. New York: Alfred A. Knopf.

Cherett, J. M., ed. 1989. *Ecological Concepts: The Contribution of Ecology to an Understanding of the Natural World.* Oxford: Blackwell.

Clarke, A. 1990. A social worlds research adventure: The case of reproductive science. In S. E. Cozzens and T. F. Gieryn, eds., *Theories of Science in Society,* 15–42. Bloomington: Indiana University Press.

———. 1991. Social worlds/arenas theory as organizational theory. In D. R. Maines, ed., *Social Organization and Social Process: Essays in Honor of Anselm Strauss,* 119–58. New York: Aldine de Gruyter.

Clarke, A., and J. Fujimura, eds. 1992a. *The Right Tools for the Job: At Work in Twentieth-Century Life Sciences.* Princeton, NJ: Princeton University Press.

———. 1992b. What tools? Which jobs? Why right? In A. Clarke and J. Fujimura, eds., *The Right Tools for the Job: At Work in Twentieth-Century Life Sciences,* 3–44. Princeton, NJ: Princeton University Press.

Clarke, G. L. 1946. Dynamics of production in a marine area. *Ecological Monographs* 16 (4): 321–35.

———. 1954 [1965]. *Elements of Ecology.* New York: Wiley.

Clements, F. E. 1916. *Plant Succession: An Analysis of the Development of Vegetation.* Washington, DC: Carnegie Institution of Washington.

Clements, F. E., and V. Shelford. 1939. *Bio-ecology.* New York: John Wiley and Sons.

Cody, M. L., and J. M. Diamond, eds. 1975. *Ecology and the Evolution of Communities.* Cambridge, MA: Harvard University Press.

Coker, R. E. 1939. Some philosophical reflections of a biologist. *Science Monthly* 48:61–68, 121–29.

Collins, H. M., ed. 1981a. Knowledge and contingency. Special issue, *Social Studies of Science* 11:3–158.

———. 1981b. Stages in the empirical programme of relativism. *Social Studies of Science* 11:3–10.

———. 1984. Researching spoonbending: Concepts and practice of participatory fieldwork. In C. Bell and H. Roberts, eds., *Social Researching: Politics, Problems, Practice,* 54–69. London: Routledge & Kegan Paul.

Commoner, B. 1963. *Science and Survival.* New York: Viking Press.

———. 1971. *The Closing Circle.* New York: Knopf.

Cook, R. E. 1977. Raymond Lindeman and the trophic-dynamic concept in ecology. *Science* 198:22–26.

Cronon, W. 1983. *Changes in the Land.* New York: Hill & Wang.

Cross, S. J., and W. R. Albury. 1987. Walter B. Cannon, L. J. Henderson, and the Organic Analogy. *Osiris* 3:165–92.

Dalby, D., and R. J. Harrison Church. 1973. *Drought in Africa.* London: University of London, School of African and Oriental Studies.

Danziger, K. 1990. Generative metaphor and the history of psychological discourse. In D. E. Leary, ed., *Metaphors in the History of Psychology,* 331–56. Cambridge: Cambridge University Press.

Dawe, A. 1978. Theories of social action. In T. Bottomore and R. Nisbet, eds., *A History of Sociological Analysis,* 362–417. New York: Basic Books.

DeAngelis, D. L. 1975. Stability and connectance in food web models. *Ecology* 56:238–43.

———. 1992. *Dynamics of Nutrient Cycling and Food Webs.* London: Chapman and Hall.

DeAngelis, D. L., and L. J. Gross, eds. 1992. *Populations and Communities: An Individual-Based Perspective.* New York: Chapman and Hall.

DeAngelis, D. L., and J. C. Waterhouse. 1987. Equilibrium and non-equilibrium concepts in ecological models. *Ecological Monographs* 57:1–21.

Dennis, M. A. 1997. Historiography of Science: An American Perspective. In J. Krige and D. Pestre, eds., *Science in the Twentieth Century,* 1–26. Amsterdam: Harwood Academic Publishers.

Dervin, B. 1999. Chaos, order, and sense-making: A proposed theory for information design. In R. Jacobson, ed., *Information Design,* 35–57. Cambridge, MA: MIT Press.

Diamond, J. 1975. Assembly of species communities. In M. L. Cody and J. Diamond, eds., *Ecology and Evolution of Communities,* 342–444. Cambridge, MA: Belknap Press of Harvard University Press.

Dickson, D. 1984. *The New Politics of Science.* New York: Pantheon.

Dirzo, R., and J. Sarukhán, eds. 1984. *Perspectives on Plant Population Ecology.* Sunderland, MA: Sinauer Associates.

Downey, G., and J. Dumit, eds. 1997. *Cyborgs and Citadels: Anthropological Interventions in Emerging Sciences and Technologies.* Santa Fe, NM: School of American Research Press.

Drake, J. A. 1990. The mechanics of community assembly and succession. *Journal of Theoretical Biology* 147:213–33.

———. 1991. Community-assembly mechanics and the structure of an experimental species ensemble. *American Naturalist* 137:1–26.

Drake, J. A., C. R. Zimmerman, T. Purucker, and C. Rojo. 1999. On the nature of the assembly trajectory. In E. Weiher and P. Keddy, eds., *Ecological Assembly Rules: Perspectives, Advances, Retreats,* 233–50. Cambridge: Cambridge University Press.

Durham, W. D. 1979. *Scarcity and survival in Central America: Ecological origins of the soccer war.* Stanford, CA: Stanford University Press.

Ebata, T. 1997. Adaptive Management References. www.for.gov.bc.ca/hfp/amhome/annobib/ambib.htm (viewed 6/14/00).

Edwards, P. N. 2005. *The World in a Machine: Computer Models, Data Networks, and Global Atmospheric Politics.* Cambridge, MA: MIT Press. Forthcoming.

Egerton, F. N. 1973. Changing concepts of the balance of nature. *Quarterly Review of Biology* 48:322–50.

Eglash, R. 1999. *African Fractals: Modern Computing and Indigenous Design.* New Brunswick, NJ: Rutgers University Press.

Ehrlich, P. R., A. H. Ehrlich, and G. C. Daily. 1995. *The Stork and the Plow: The Equity Answer to the Human Dilemma.* New York: G. P. Putnam's Sons.

Elbow, P. 1981. *Writing with Power.* New York: Oxford University Press.

Elkins, J. 1999. *The Domain of Images.* Ithaca, NY: Cornell University Press.

Elster, J. 1978. *Logic and Society: Contradiction and Possible Worlds.* New York: Wiley.

Faith, D. P., P. R. Minchin, and L. Belbin. 1987. Compositional dissimilarity as a robust measure of ecological distance. *Vegetatio* 69:57–68.

Fausto-Sterling, A. 2000. *Sexing the Body: Gender, Politics and the Construction of Sexuality.* New York: Basic Books.

Feeny, D., F. Berkes, B. McCay, and J. Acheson. 1990. The tragedy of the commons: Twenty-two years later. *Human Ecology* 18 (1): 1–19.

Ferguson, J., A. Smith, and P. Taylor. 1978. *Economic Aspects of the Use of Water Resources in the Kerang Region.* Report no. 1 to the Ministry of Water Resources. Melbourne, Australia: Institute of Applied Economic and Social Research.

———. 1979. *Economic Aspects of the Use of Water Resources in the Kerang Region.* Technical paper no. 11. Melbourne, Australia: Institute of Applied Economic and Social Research.

Fish, S. 1989. Anti-foundationalism, theory hope, and the teaching of composition. In *Doing What Comes Naturally: Change, Rhetoric, and the Practice of Theory in Literary and Legal Studies,* 343–55. Durham, NC: Duke University Press.

Fogel, R. W., and S. L. Engerman. 1969. A model for the explanation of industrial expansion during the nineteenth century: With an application to the American iron industry. *Journal of Political Economy* 77:306–28.

Forrester, J. W. 1961. *Industrial Dynamics.* Cambridge, MA: MIT Press.

———. 1969. *Urban Dynamics.* Cambridge, MA: MIT Press.

Foucault, M. 1981. Questions of method: An interview with Michel Foucault. *I&C* 8:3–14.

Frank, L. J. 1945. Research after the War. *Science* 101:433–34.

———. 1948. Foreword. *Annals of the New York Academy of Sciences* 50:189–96.

Frantz, C. 1981. Settlement and migration among pastoral Fulbe in Nigeria and Cameroun. In P. C. Salzman, ed., *Contemporary Nomadic and Pastoral Peoples: Africa and Latin America,* 57–94. Williamsburg, VA: College of William and Mary, Department of Anthropology.

Fujimura, J. 1987. Constructing "do-able" problems in cancer research: Articulating alignment. *Social Studies of Science* 17:257–93.

———. 1988. The molecular biology bandwagon in cancer research: Where social worlds meet. *Social Problems* 35:261–83.

Fuller, S. 2000. *Thomas Kuhn: A Philosophical History for Our Times.* Chicago: University of Chicago Press.

Galaty, J. G., and D. L. Johnson, eds. 1990. *The World of Pastoralism: Herding Systems in Comparative Perspective.* New York: Guilford.

García-Barrios, R., and L. García-Barrios. 1990. Environmental and technological degradation in peasant agriculture: A consequence of development in Mexico. *World Development* 18 (11): 1569–85.

Gardner, M. R., and W. R. Ashby. 1970. Connectance of large, dynamical (cybernetic) systems: Critical values for stability. *Nature* 254:137–39.

Gergen, K. J. 1990. Metaphor, metatheory, and the social world. In D. E. Leary, ed., *Metaphors in the History of Psychology,* 267–99. Cambridge: Cambridge University Press.

Gibbs, L. M. 1982. *Love Canal: My Story.* Albany: State University of New York Press.

Giddens, A. 1981. Agency, institution, and time-space analysis. In K. Knorr-Cetina and A. Cicourel, eds., *Advances in Social Theory and Methodology,* 161–74. Boston: Routledge & Kegan Paul.

Giddens, A., and J. Turner, eds. 1987. *Social Theory Today.* Stanford: Stanford University Press.

Gieryn, T. F. 1999. *Cultural Boundaries of Science: Credibility on the Line.* Chicago: University of Chicago Press.

Gilbert, S. F. 1988. Cellular politics. In R. Rainger, K. Benson, and J. Maienschein, eds., *The American Development of Biology,* 311–45. Philadelphia: University of Pennsylvania Press.

Gilpin, M., and I. Hanski, eds. 1991. *Metapopulation Dynamics: Empirical and Theoretical Investigations.* London: Academic Press.

Ginzburg, L. R., H. R. Akcakaya, and J. Kim. 1988. Evolution of community structure: Competition. *Journal of Theoretical Biology* 133:513–23.

Giri, J. 1976. An analysis and synthesis of long-term development: Strategies for the Sahel. *Organization for Economic Co-operation and Development* (March).

Givnish, T. J., ed. 1986. *On the Economy of Plant Form and Function.* Cambridge: Cambridge University Press.

Glantz, M. H., ed. 1976. *Politics of natural disaster: The case of the Sahel drought.* New York: Praeger.

Glantz, M., J. Robinson, and M. E. Krenz. 1980. Recent assessments. In R. W. Kates, J. H. Ausubel, and M. Berberian, eds., *Climate Impact Assessment,* 565–98. New York: John Wiley & Sons.

Gleason, H. 1926. The individualistic concept of the plant association. *Bulletin of the Torrey Botanical Club* 53:1–20.

———. 1927. Further views on the succession concept. *Ecology* 8:299–326.

Glimcher, P. W. 2003. *Decisions, Uncertainty, and the Brain: The Science of Neuro-economics.* Cambridge, MA: MIT Press.

Göbber, F., and F. F. Seelig. 1975. Conditions for the application of the steady-state approximation to systems of differential equations. *Journal of Mathematical Biology* 2:79–86.

Goldblatt, D. 1996. *Social Theory and the Environment.* Oxford: Polity Press.

Golinski, J. 1990. The theory of practice and the practice of theory: Sociological approaches in the history of science. *Isis* 81:492–505.

Goodman, D. 1975. The theory of diversity-stability relationships in ecology. *Quarterly Review of Biology* 50:237–66.

Gould, P., and F. R. Pitts, eds. 2002. *Geographical Voices: Fourteen Autobiographical Essays.* Syracuse, NY: Syracuse University Press.

Gray, A. J., M. J. Crawley, and P. J. Edwards, eds. 1987. *Colonization, Succession and Stability.* 26th Symposium of the British Ecological Society. Oxford: Blackwell.

Greenwood, D. J., and M. Levin. 1998. *Introduction to Action Research: Social Research for Social Change.* Thousand Oaks, CA: Sage.

Griffith, J. L., and M. E. Griffith. 1994. *The Body Speaks: Therapeutic Dialogues for Mind-Body Problems.* New York: Basic Books.

Gunderson, L. H., C. S. Holling, and S. S. Light, eds. 1995a. *Barriers and Bridges to the Renewal of Ecosystems and Institutions.* New York: Columbia University Press.

———. 1995b. Barriers broken and bridges built: A synthesis. In L. H. Gunderson, C. S. Holling and S. S. Light, eds., *Barriers and Bridges to the Renewal of Ecosystems and Institutions,* 489–532. New York: Columbia University Press.

Guyer, J. I. 1988. Multiplication of labor: Historical methods in the study of gender and agricultural change in modern Africa. *Current Anthropology* 29 (2): 247–72.

Hacking, I. 1983. *Representing and Intervening.* Cambridge: Cambridge University Press.

———. 1999. *The Social Construction of What?* Cambridge, MA: Harvard University Press.

Hagen, J. B. 1989. Research perspectives and the anomalous status of modern ecology. *Biology and Philosophy* 4:433–55.

———. 1992. *The Entangled Bank: The Origins of Ecosystem Ecology.* New Brunswick, NJ: Rutgers University Press.

Haila, Y. 1988. The multiple faces of ecological theory and data. *Oikos* 53:408–11.

Haila, Y., and O. Järvinen. 1990. Northern conifer forests and their bird species assemblages. In A. Keast, ed., *Biogeography and Ecology of Forest Bird Communities,* 61–85. The Hague: SPB Academic Publishing.

Haila, Y., and R. Levins. 1992. *Humanity and Nature: Ecology, Science and Society.* London: Pluto Press.

Hall, C. A., ed. 1995. *Maximum Power: The Ideas and Applications of H. T. Odum.* Niwot, CO: University Press of Colorado.

Hall, C. A., and D. L. DeAngelis. 1985. Models in ecology: Paradigms found or paradigms lost? *Bulletin of the Ecological Society of America* 66:339–45.

Haraway, D. J. 1981–1982. High cost of information in post-World War II evolutionary biology: Ergonomics, semiotics, and the sociobiology of communication systems. *Philosophical Forum* 13 (2–3): 244–79.

———. 1983. Signs of dominance: From a physiology to a cybernetics of primate society. *Studies in History of Biology* 6:129–219.

————. 1984–1985. Teddy bear patriarchy: Taxidermy in the Garden of Eden, New York City, 1908–1936. *Social Text* 11:20–64.

————. 1988. Situated knowledges: The science question in feminism and the privilege of partial perspective. *Feminist Studies* 14 (3): 575–99.

————. 1989. Teddy bear patriarchy: Taxidermy in the garden of Eden, New York City, 1908–1936. In *Primate Visions: Gender, Race, and Nature in the World of Modern Sciences,* 26–58. New York: Routledge.

Haraway, D. J., and T. N. Goodeve. 2000. *How Like a Leaf.* New York: Routledge.

Hardin, G. 1968. The tragedy of the commons. *Science* 162:1243–48.

Harmsworth, J. 1984. Small-scale projects and Sahel nomads. *Cultural Survival Quarterly* 8:59–64.

Harré, R., and E. H. Madden. 1975. *Causal Powers: A Theory of Natural Necessity.* Oxford: Basil Blackwell.

Hart, H. L. A., and A. M. Honoré. 1959. *Causation in the Law.* Oxford: Clarendon Press.

Harvey, D. 1969. *Explanation in Geography.* London: Edward Arnold.

————. 1989. *The Condition of Postmodernity: An Enquiry into the Origins of Cultural Change.* Oxford: Blackwell.

————. 1993. The nature of the environment: The dialectics of social and environmental change. *The Socialist Register* 1993:1–51.

————. 1995. Militant particularism and global ambition: The conceptual politics of place, space, and environment in the work of Raymond Williams. *Social Text* 42:69–98.

Harvey, M. J. 1974. Letter to Congressman Diggs, April 23. Private collection of Michael Glantz, National Center for Atmospheric Research.

Harwood, J. 2004. National differences in academic cultures: Science in Germany and the United States between the world wars. In C. Charle, J. Schriewer, and P. Wagner, eds., *Transnational Intellectual Networks: Forms of Academic Knowledge and the Search for Cultural Identities,* 53–79. Frankfurt a.M: Campus.

Hastings, A., and S. Harrison. 1994. Metapopulation dynamics and genetics. *Annual Review of Ecology and Systematics* 25:167–88.

Hawthorn, G. 1991. *Plausible Worlds.* Cambridge: Cambridge University Press.

Hayter, T., and D. Harvey, eds. 1993. *The Factory and the City: The Story of the Cowley Automobile Workers in Oxford.* Brighton: Mansell.

Heims, S. 1980. *John von Neumann and Norbert Wiener.* Cambridge, MA: MIT Press.

————. 1991. *The Cybernetics Group: Constructing a Social Science for Postwar America.* Cambridge, MA: MIT Press.

Helmreich, S. G. 1999. Digitizing development: Balinese water temples, complexity and the politics of simulation. *Critique of Anthropology* 19 (3): 249–65.

Hendriks-Jansen, H. 1996. *Catching Ourselves in the Act.* Cambridge, MA: MIT Press.

Herrera, C. M. 1986. Vertebrate-dispersed plants: Why they don't behave the way they should. In A. Estrada and T. H. Fleming, eds., *Frugivores and Seed Dispersal,* 5–18. Dordrecht: Dr. W Junk Publishers.

Hess, D. J. 1997. *Science Studies: An Advanced Introduction*. New York: New York University Press.

Hesse, M. 1966. *Models and Analogies in Science*. Notre Dame, IN: University of Notre Dame Press.

Hilborn, R., and S. Stearns. 1982. On inference in ecology and evolutionary biology: The problem of multiple causes. *Acta Biotheoretica* 32:145–64.

Hobson, L. 1974. USAID internal memorandum. Private collection of Michael Glantz, National Center for Atmospheric Research.

Hofrichter, R., ed. 2000. *Reclaiming the Environmental Debate: The Politics of Health in a Toxic Culture*. Cambridge, MA: MIT Press.

Holle, O., and M. Knell. 1996. A tragedy of the commons game, presented at the SASE-Conference on Computer Modelling for Socio-economic Systems. olymp.wu-wien.ac.at/usr/ai/mitloehn/commons/ (viewed 10/12/00).

Holling, C. S. 1978. *Adaptive Environmental Assessment and Management*. London: John Wiley.

———. 1995. What barriers? What bridges? In L. H. Gunderson, C. S. Holling, and S. S. Light, eds., *Barriers and Bridges to the Renewal of Ecosystems and Institutions*, 3–36. New York: Columbia University Press.

Holt, R. D. 1977. Predation, apparent competition, and the structure of prey communities. *Theoretical Population Biology* 12:197–229.

Horowitz, M., ed. 1976. *Colloquium on the Effects of Drought on the Productive Strategies of Sudano-Sahelian Herders and Farmers: Implications for Development*. Binghamton, NY: Institute for Development Anthropology.

Horowitz, M. M., and P. D. Little. 1987. African pastoralism and poverty: Some implications for drought and famine. In M. Glantz, ed., *Drought and hunger in Africa: Denying famine a future*, 59–82. Cambridge: Cambridge University Press.

Hughes, E. C. 1971. *The Sociological Eye*. Chicago: Aldine Atherton.

Humphries, J. 1990. Enclosures, common rights, and women: The proletarianization of families in the late eighteenth and early nineteenth centuries. *Journal of Economic History* 50 (1): 17–42.

Huston, M. A., L. W. Aarssen, M. P. Austin, B. S. Cade, J. D. Fridley, E. Garnier, J. P. Grime, J. Hodgson, and W. K. Lauenroth. 2000. No consistent effect of plant diversity on productivity. *Science* 290:1255 [www.sciencemag.org/cgi/content/full/289/5483/1255a].

Huston, M. A., D. DeAngelis, and W. Post. 1988. From individuals to ecosystems: A new approach to ecological theory. *Bioscience* 38:682–91.

Hutchinson, G. E. 1940. Review: *Bio-ecology*. *Ecology* 21:267–68.

———. 1942. Addendum to R. Lindeman, "The trophic-dynamic aspect of ecology." *Ecology* 23:417.

———. 1945. Marginalia. *American Scientist* 33:262–69.

———. 1946. Social theory and social engineering. *Science* 104:166–67.

———. 1948. Circular causal systems in ecology. *Annals of the New York Academy of Sciences* 50:221–23,236–46.

———. 1949. Marginalia: Review of Wiener's *Cybernetics*. *American Scientist* 37:267.

———. 1978. *An Introduction to Population Ecology*. New Haven, CT: Yale University Press.

Hymes, D. 1974. Traditions and paradigms. In D. Hymes, ed., *Studies in the History of Linguistics: Traditions and Paradigms,* 1–38. Bloomington: University of Indiana Press.

Illich, I. 1973. *Tools for Conviviality*. New York: Harper & Row.

———. 1976. *Medical Nemesis: The Expropriation of Health*. New York: Pantheon Books.

INTECOL. 1974. *Proceedings of the 1st International Congress of Ecology*. Wageningen: Centre for Agricultural Publications and Documents.

Isaacs, W. 1999. *Dialogue and the Art of Thinking Together*. New York: Currency.

Jacobson, M. C., R. J. Charlson, H. Rodhe, G. H. Orians, M. F. Jacobson, and H. Rodhe, eds. 2000. *Earth System Science: From Biogeochemical Cycles to Global Change*. San Diego, CA: Academic Press.

Johnson, G. 2000. First cells, then species, now the Web. *The New York Times* (26 December).

Jones, C. A., and P. L. Galison, eds. 1998. *Picturing Science: Producing Art*. New York: Routledge.

Kaiser, J. 2000a. Ecologists on a mission to save the world. *Science* 287 (18 February): 1188–92.

———. 2000b. Rift over biodiversity divides ecologists. *Science* (25 August): 1282–83.

Kargon, R. 1983. The future of American science: An historical perspective. In M. E. Kahn, ed., *The Future of American Democracy*. Philadelphia: Temple University Press.

Kargon, R., and E. Hodes. 1985. Karl Compton, Isaiah Bowman, and the politics of science in the Great Depression. *Isis* 76:301–18.

Kassas, M. 1970. Desertification versus potential for recovery in circum-Saharan territories. In H. E. Dregne, ed., *Arid Lands in Transition,* 123–42. Washington, DC: American Association for the Advancement of Science.

Keddy, P., and E. Weiher. 1999. The scope and goals of research on assembly rules. In E. Weiher and P. Keddy, eds., *Ecological Assembly Rules: Perspectives, Advances, Retreats,* 1–20. Cambridge: Cambridge University Press.

Keller, E. F. 1988. Demarcating public from private values. *Journal of the History of Biology* 21 (2): 195–211.

———. 1992. Critical silences in scientific discourse: Problems of form and re-form. In *Secrets of Life, Secrets of Death: Essays on Language, Gender and Science,* 73–92. New York: Routledge.

———. 2001. Making a difference: Feminist movement and feminist critiques of science. In A. Creager, E. Lunbeck, and L. Schiebinger, eds., *Feminism in Twentieth-Century Science, Technology, and Medicine,* 98–109. Chicago: University of Chicago Press.

Kelt, D. A., and J. H. Brown. 1999. Community structure and assembly rules: Confronting conceptual and statistical issues with data on desert rodents. In E. Weiher and P. Keddy, eds., *Ecological Assembly Rules: Perspectives, Advances, Retreats*, 75–107. Cambridge: Cambridge University Press.

Kingsland, S. 1995. *Modeling Nature: Episodes in the History of Population Ecology.* 2nd. ed. Chicago: University of Chicago Press.

———. 2005. New frontiers. In *The Evolution of Ecology: Science and American Society.* Baltimore: Johns Hopkins University Press. Forthcoming.

Kinzig, A. 2000. Nature and society: An imperative for Integrated Environmental Research, http://lsweb.la.asu.edu/akinzig/report.htm (viewed 11/16/00).

Kitcher, P. 1993. *The Advancement of Science: Science without Legend, Objectivity without Illusions.* New York: Oxford University Press.

Klopateck, J. M., and R. H. Gardner, eds. 1999. *Landscape Ecological Analysis: Issues and Applications.* New York: Springer-Verlag.

Knorr-Cetina, K. 1993. Strong constructivism—from a sociologist's point of view: A personal addendum to Sismondo's paper. *Social Studies of Science* 23 (3): 555–63.

Kondo, D. K. 1990. The eye/I. In *Crafting Selves: Power, Gender, and Discourses of Identity in a Japanese Workplace*, 3–48. Chicago: University of Chicago Press.

Krieger, N. 1994. Epidemiology and the web of causation: Has anyone seen the spider? *Social Science & Medicine* 39 (7): 887–903.

Kuhn, T. 1970. *The Structure of Scientific Revolutions.* Chicago: University of Chicago Press.

———. 1977. The essential tension: Tradition and innovation in scientific research. In *The essential tension: Selected studies in scientific tradition and change*, 225–39. Chicago: University of Chicago Press.

Kwa, C. 1989. *Mimicking Nature: The Development of Systems Ecology in the United States, 1950–1975.* Amsterdam: University of Amsterdam.

Lakoff, G. 1993. The contemporary theory of metaphor. In A. Ortony, ed., *Metaphor and Thought*, 202–51. Cambridge: Cambridge University Press.

Lakoff, G., and M. Johnson. 1980. *Metaphors We Live By.* Chicago: University of Chicago Press.

Lamb, H. H. 1973. Is the climate changing? *UNESCO Courier* 26 (August–September): 17–20.

Latour, B. 1987. *Science in Action: How to Follow Scientists and Engineers through Society.* Milton Keynes: Open University Press.

———. 1988a. Mixing humans and non-humans together: The sociology of a door-closer. *Social Problems* 35 (3): 298–310.

———. 1988b. The politics of explanation: An alternative. In S. Woolgar, ed., *Knowledge and Reflexivity: New Frontiers in the Sociology of Knowledge*, 155–76. London: Sage.

———. 1994. Les objets ont-ils une histoire? Recontre de Pasteur et de Whitehead dans un bain d'acide lactique. In I. Stengers, ed., *L'effet Whitehead*, 197–217. Paris: Vrin.

———. 1999. *Pandora's Hope: Essays on the Reality of Science Studies.* Cambridge, MA: Harvard University Press.

———. 2004. *Politics of Nature: How to Bring the Sciences into Democracy.* Cambridge, MA: Harvard University Press.

Law, J. 1986. On the methods of long-distance control: Vessels, navigation and the Portuguese route to India. In J. Law, ed., *Power, Action, Belief,* 234–63. London: Routledge & Kegan Paul.

———. 1987. Technology and heterogeneous engineering: The case of Portuguese expansion. In W. E. Bijker, T. P. Hughes, and T. J. Pinch, eds., *The Social Construction of Technological Systems: New Directions in the Sociology and History of Technology,* 111–34. Cambridge, MA: MIT Press.

Law, J., and A. Mol, eds. 2002. *Complexities: Social Studies of Knowledge Practices.* Durham, NC: Duke University Press.

Lawlor, L. R. 1978. A comment on randomly constructed ecosystem models. *American Naturalist* 112:445–47.

———. 1979. Direct and indirect effects of *n*-species competition. *Oecologia* 43:355–64.

———. 1980. Structure and stability in natural and randomly constructed competitive communities. *American Naturalist* 116:394–408.

Lawlor, L. R., and J. Maynard-Smith. 1976. The coevolution and stability of competing species. *American Naturalist* 110:79–99.

Lee, K. 1993. *Compass and Gyroscope: Integrating Science and Politics for the Environment.* Washington, DC: Island Press.

Levidow, L., ed. 1986. *Radical Science.* London: Free Association Books.

Levidow, L., and B. Young, eds. 1981. *Science, Technology and the Labour Process: Marxist Studies.* London: CSE Books.

Levin, S. 1980. Mathematics, ecology, ornithology. *Auk* 97:422–25.

———. 2000. *Fragile Dominion: Complexity and the Commons.* Boulder, CO: Perseus Books.

Levine, S. H. 1976. Competitive interactions in ecosystems. *American Naturalist* 110:903–10.

Levins, R. 1966. The strategy of model building in population biology. *American Scientist* 54:421–31.

———. 1993. A response to Orzack and Sober: Formal analysis and the fluidity of science. *Quarterly Review of Biology* 68 (4): 547–55.

Levins, R., and R. Lewontin. 1985. *The Dialectical Biologist.* Cambridge, MA: Harvard University Press.

Lewontin, R. C. 1974. The analysis of variance and the analysis of causes. *American Journal of Human Genetics* 26:400–411.

Life. 1950. U.S. science holds its biggest powwow. (9 January): 20.

Lindeman, R. 1942. The trophic-dynamic aspect of ecology. *Ecology* 23:399–417.

Little, P. 1985. Social differentiation and pastoralist sedentarization in Northern Kenya. *Africa* 55:243–61.

―――. 1987. Land use conflicts in the agricultural/pastoral borderlands: The case of Kenya. In P. Little, M. Horowitz, and A. Nyerges, eds., *Lands at Risk in the Third World: Local Level Perspectives*, 195–212. Boulder, CO: Westview.

Lloyd, C. 1986. *Explanation in Social History*. Oxford: Basil Blackwell.

Lloyd, E. A. 1987. Confirmation of ecological and evolutionary models. *Biology & Philosophy* 2:277–93.

Lomnicki, A. 1980. Regulation of plant density due to individual differences and patchy environment. *Oikos* 35:185–93.

Longino, H. 1990. *Science as Social Knowledge*. Princeton, NJ: Princeton University Press.

Lotka, A. J. 1922. Contributions to the energetics of evolution. *Proceedings of the National Academy of Sciences of the United States of America* 8:147–50.

―――. 1925. *Elements of Physical Biology*. Baltimore: Williams and Wilkins.

―――. 1956. *Elements of Mathematical Biology*. New York: Dover.

Lyman, P. 1974. Review of the MIT Interim Report on the Sahel-Sudan Area. USAID memorandum, March 25. Private collection of Michael Glantz, National Center for Atmospheric Research.

―――. 1976. Evaluation of the MIT contract. USAID Memorandum, February. Private collection of Michael Glantz, National Center for Atmospheric Research.

Lynch, M. 1991. Science in the age of mechanical reproduction: Moral and epistemic relations between diagrams and photographs. *Biology & Philosophy* 6:205–26.

―――. 1993. *Scientific Practice and Ordinary Action: Ethnomethodology and Social Studies of Science*. Cambridge: Cambridge University Press.

―――. 2001. Visualization, representation. In N. J. Smelser and P. B. Baltes, eds., *International Encyclopedia of the Social and Behavioral Sciences*. New York: Pergamon.

Lynch, M., and S. Woolgar, eds. 1990. *Representation in Scientific Practice*. Cambridge, MA: MIT Press.

Lynch, W. T. 1989. Arguments for a non-Whiggish hindsight: Counterfactuals and the sociology of knowledge. *Social Epistemology* 3 (4): 361–65.

―――. 2003. Beyond cold war paradigms for science and democracy: Review of Steve Fuller, *Thomas Kuhn: A Philosophical History for Our Times. Minerva* 41 (4): 365–79.

MacArthur, R. H. 1955. Fluctuations of animal populations, and a measure of community stability. *Ecology* 35:533–36.

―――. 1972. *Geographical Ecology*. New York: Harper and Row.

Mackie, J. L. 1965. Causes and conditionals. *American Philosophical Quarterly* 2:245–65.

Margalef, R. 1958. Information theory in ecology. *General Systems* 3:36–71.

Markus, G. 1986. *Language and Production*. Dordrecht: D. Reidel.

Martin, B., ed. 1996a. *Confronting the Experts*. Albany: State University of New York Press.

————. 1996b. Sticking a needle into science: The case of polio vaccine and the origins of AIDS. *Social Studies of Science* 26 (2): 245–76.

Martin, B., and E. Richards. 1994. Scientific knowledge, controversy, and public decision making. In S. Jasanoff, G. Markel, J. Peterson, and T. Pinch, eds., *Handbook of Science and Technology Studies,* 506–26. Beverly Hills, CA: Sage.

Martinez, N. D. 1991. Artifacts or attributes? Effects of resolution on the Little Rock Lake food web. *Ecological Monographs* 61:367–92.

May, R. M. 1972. Will a large complex system be stable? *Nature* 238:413–14.

————. 1973. *Stability and Complexity in Model Ecosystems.* Princeton, NJ: Princeton University Press.

————. 1984. An overview: Real and apparent patterns in community structure. In D. R. Strong, D. Simberloff, L. G. Abele, and A. B. Thistle, eds., *Ecological Communities: Conceptual Issues and the Evidence,* 3–16. Princeton, NJ: Princeton University Press.

————. 2000. Relation between diversity and stability, in the real world. *Science* 290 (27 October): 714–15.

McCabe, J. T., and J. E. Ellis. 1987. Beating the odds in arid Africa. *Natural History* 1:33–40.

McCay, B. J. 1992. Everyone's concern, whose responsibility? The problem of the commons. In S. Ortiz and S. Lees, eds., *Understanding Economic Processes,* 189–210. Lanham, MD: University Press of America.

McCay, B. J, and S. Jentoft. 1998. Market or community failure? Critical perspectives on common property research. *Human Organization* 57 (1): 21–29.

McCulloch, W. 1948. A recapitulation of the theory, with a forecast of several extensions. *Annals of the New York Academy of Sciences* 50:247–58.

McKean, M., and E. Ostrom. 1995. Common property regimes in the forest: Just a relic from the past? *Unasylva* 46:3–15.

McLaughlin, P. 1989. Obstacles to a new sociology of agriculture: The persistence of essentialism. Working Paper, Department of Rural Sociology, Cornell University.

McMurtrie, R. E. 1975. Determinants of stability of large randomly connected systems. *Journal of Theoretical Biology* 50:1–11.

McNaughton, S. J. 1977. Diversity and stability of ecological communities: A comment on the role of empiricism in ecology. *American Naturalist* 111:515–25.

Meadows, D., D. Meadows, J. Randers, and W. W. Behrens. 1972. *The Limits to Growth.* New York: Universe Books.

Meffe, G. K., A. H. Ehrlich, and D. Ehrenfeld. 1993. Human population control: The missing agenda. *Conservation Biology* 7 (1): 1–3.

Mertz, D. B., and D. E. McCauley. 1980. The domain of laboratory ecology. *Synthese* 43:95–110.

Miller, R. W. 1991. Fact and method in the social sciences. In R. Boyd, P. Gasper, and J. D. Trout, eds., *The Philosophy of Science,* 743–62. Cambridge, MA: MIT Press.

Minchin, P. R. 1987. An evaluation of the relative robustness of techniques for ecological ordination. *Vegetatio* 69:89–107.

Mitchell, R. 1997. A tragedy of the commons game. darkwing.uoregon.edu/~rmitchel/tragedy/ (viewed 10/12/00).

Mitchell, T. 1990. Everyday metaphors of power. *Theory and Society* 19:545–77.

Mitchell, W. J. T. 1986. *Iconology*. Chicago: University of Chicago Press.

Mitman, G. 1988. From population to society: The cooperative metaphors of W. C. Allee and A. E. Emerson. *Journal of the History of Biology* 21:173–94.

———. 1992. *The State of Nature: Ecology, Community, and American Social Thought, 1900–1950*. Chicago: University of Chicago Press.

———. 1994. Defining the organism in the welfare state: The politics of individuality in American culture, 1890–1950. *Social Sciences Yearbook* 18:249–80.

———. 1999. *Reel Nature: America's Romance with Wildlife on Film*. Cambridge, MA: Harvard University Press.

Moke, S. 1994. Challenging the conventional wisdom. *Research & Creative Activity (Indiana University)* 16 (3): 15–18.

Moore, J. 1997. Wallace's Malthusian moment: The common context revisited. In B. Lightman, ed., *Victorian Science in Context,* 290–311. Chicago: University of Chicago Press.

Morentz, J. W. 1980. Communication in the Sahel drought: Comparing the mass media with other channels of international communication. In *Disasters and the Mass Media,* 158–86. Washington, DC: National Academy of Sciences.

Münch, R. 1987. Parsonian theory today: In search of a new synthesis. In A. Giddens and J. Turner, eds., *Social Theory Today,* 116–55. Stanford: Stanford University Press.

NAS. *See* National Academy of Sciences.

National Academy of Sciences. 1974. *United States Participation in the International Biological Program*. Washington, DC: National Academy of Sciences.

———. 1975a. Arid lands of sub-Saharan Africa: Staff progress report, September 1973–June 1974. Washington, DC: National Academy of Sciences.

———. 1975b. Arid lands of sub-Saharan Africa: Staff final report, July 1974–December 1974. Washington, DC: National Academy of Sciences.

Nee, S. 1990. Community construction. *Trends in Ecology and Evolution* 5:337–40.

Nelkin, D., ed. 1984. *Controversy: Politics of Technical Decisions*. Beverly Hills, CA: Sage.

Newsweek. 1975. Odum's law. January 13.

Novak, J. D. 1990. Concept mapping: A useful tool for science education. *Journal of Research in Science Teaching* 27 (10): 937–49.

Nunney, L. 1980. The stability of complex model ecosystems. *American Naturalist* 115:639–49.

Odling-Smee, F. J., K. N. Laland, and M. W. Feldman. 2003. *Niche Construction: The Neglected Process*. Princeton, NJ: Princeton University Press.

Odum, E. P. 1953/1959/1971. *Fundamentals of Ecology*. Philadelphia: Saunders.

———. 1968. Energy flow in ecosystems: A historical review. *American Zoologist* 8:11–18.

Odum, H. T. 1950. The biogeochemistry of strontium. Ph.D. dissertation, Yale University, New Haven, CT.

———. 1951. The stability of the world strontium cycle. *Science* 114:407–11.

———. 1955. Trophic structure and productivity of Silver Springs, Florida. *Ecological Monographs* 27:55–112.

———. 1956a. Efficiencies, size of organisms, and community structure. *Ecology* 37:592–97.

———. 1956b. Primary production in flowing waters. *Limnology and Oceanography* 1:102–17.

———. 1960. Ecological potential and analogue circuits for the ecosystem. *American Scientist* 48:1–8.

———. 1962a. Ecological tools and their use; man and the ecosystem. *Bulletin of the Connecticut Agricultural Station* 652:57–75.

———. 1962b. The use of a network simulator to synthesize systems and develop analogous theory: The ecosystem example. In H. L. Lucas, ed., *Proceedings of the Cullowhee Conference Training in Biomathematics,* 291–97. Raleigh: North Carolina State University.

———. 1967. Biological circuits and the marine systems of Texas. In T. A. Olson and F. J. Burgess, eds., *Pollution and Marine Ecology,* 99–157. New York: Interscience.

———. 1970. Summary: An emerging view of the ecological system at El Verde. In H. T. Odum and R. F. Pigeon, eds., *A Tropical Rain Forest: A Study of Irradiation and Ecology at El Verde, Puerto Rico.* Oak Ridge, TN.: U.S. Atomic Energy Commission, I-191–I-289.

———. 1971. *Environment, Power, and Society*. New York: Wiley-Interscience.

———. 1982. *Systems Ecology: An Introduction*. New York: Wiley.

———. 1986. Interview by Peter Taylor in Gainesville, Florida, January 14th. Tapes available from the archives of the Ecological Society of America.

Odum, H. T., and C. M. Hoskin. 1958. Comparative studies of the metabolism of Texas bays. In *Comparative studies on the metabolism of marine waters,* 65–96. Publication 5, Texas University Institute of Marine Science.

Odum, H. T., and E. P. Odum. 1955. Trophic structure and productivity of a windward coral reef community on Eniwetok Atoll. *Ecological Monographs* 25:291–320.

Odum, H. T., and R. C. Pinkerton. 1955. Time's speed regulator: The optimum efficiency for maximum power output in physical and biological systems. *American Scientist* 43:321–43.

Odum, H. T., M. Sell, M. Brown, J. Zuchetto, C. Swallows, J. Browder, T. Ahlstrom, and L. Peterson. 1974. Models of herbicide, mangroves, and war in South Vietnam. In National Research Council, Committee on the Effects of Herbicides in Vietnam, *The Effects of Herbicides in South Vietnam,* part B, Working Papers. Washington, DC: National Academy of Sciences.

Odum, H. W. 1927. *Man's Quest for Social Guidance*. New York: Henry Holt.

————. 1947. *Understanding Society: The Principles of Dynamic Sociology*. New York: MacMillan.

Ogburn, W. F. 1922. *Social Change with Respect to Culture and Original Nature*. New York: B. W. Huebsch.

Okoye, S. E., and P. B. Smith. 1994. Introduction. In P. B. Smith, S. E. Okoye, J. D. Wilde, and P. Deshingkar, eds., *The World at the Crossroads: Towards a Sustainable, Equitable and Liveable World*, 1–17. London: Earthscan.

Olsson, G. 2002. Glimpses. In P. Gould and F. R. Pitts, eds., *Geographical Voices: Fourteen Autobiographical Essays*, 237–68. Syracuse, NY: Syracuse University Press.

O'Neill, R. V., D. L. DeAngelis, J. B. Waide, and T. F. H. Allen. 1986. *A Hierarchical Concept of the Ecosystem*. Princeton, NJ: Princeton University Press.

Oreskes, N., K. Shrader-Frechette, and K. Belitz. 1994. Verification, validation, and confirmation of numerical models in the earth sciences. *Science* 263:641–46.

Ortony, A., ed. 1993. *Metaphor and Thought*. Cambridge: Cambridge University Press.

Orzack, S. H., and E. Sober. 1993. A critical assessment of Levins's *The strategy of model building in population biology*. *Quarterly Review of Biology* 68 (4): 533–46.

Ostrom, E. 1990. *Governing the Commons: The Evolution of Institutions for Collective Action*. New York: Cambridge University Press.

————. 1993. Design principles in long-enduring irrigation institutions. *Water Resources Research* 29 (7): 1907–12.

Oyama, S. 1985. *The Ontogeny of Information*. Cambridge: Cambridge University Press.

Oyama, S., P. Griffiths, and R. Gray, eds. 2001. *Cycles of Contingency: Developmental Systems and Evolution*. Cambridge, MA: MIT Press.

Palladino, P. 1991. Defining ecology: Ecological theories, mathematical models, and applied biology in the 1960s and 1970s. *Journal of the History of Biology* 24:223–43.

Park, P., M. Brydon-Miller, B. Hall, and T. Jackson. 1993. *Voices of Change: Participatory Research in the United States and Canada*. Westport, CT: Bergin & Garvey.

Patten, B. C. 1959. An introduction to the cybernetics of the ecosystem: The trophic dynamic aspect. *Ecology* 40:221–31.

Peet, R., and M. Watts, eds. 1996a. *Liberation Ecologies: Environment, Development, Social Movements*. London: Routledge.

————. 1996b. Liberation ecology: Development, sustainability, and environment in an age of market triumphalism. In R. Peet and M. Watts, eds., *Liberation Ecologies: Environment, Development, Social Movements*, 1–45. London: Routledge.

Pellow, D. N. 2002. *Garbage Wars: The Struggle for Environmental Justice in Chicago*. Cambridge, MA: MIT Press.

Peluso, N. 1993. Coercing conservation: The politics of state resource control. *Global Environmental Change* 3 (2): 199–217.

Peters, P. 1987. Embedded systems and rooted models: The grazing lands of Botswana and the commons debate. In B. J. McKay and J. M. Acheson, eds., *The Question of the Commons: The Culture and Ecology of Communal Resources,* 171–94. Tucson: University of Arizona Press.

———. 1996. "Who's local here?" The politics of participation in development. *Cultural Survival Quarterly* 20 (3): 22–60.

Peters, R. H. 1991. *A Critique for Ecology.* Cambridge: Cambridge University Press.

Picardi, A. C. 1973. A demographic and economic growth model for Bolivia. *Simulation* 20:109–18.

———. 1974a. Internal project memorandum to H. Findakly evaluating the project, "Large-scale interdisciplinary projects." Private collection of Michael Glantz, National Center for Atmospheric Research.

———. 1974b. Memorandum from Picardi to Godiksen, Paden, and Skinner. Private collection of Michael Glantz, National Center for Atmospheric Research.

———. 1974c. A systems analysis of pastoralism in the West African Sahel. In W. W. Seifert and N. Kamrany, eds., *A Framework for Evaluating Long-Term Strategies for the Development of the Sahel-Sudan Region,* Annex 5. Cambridge, MA: MIT, Center for Policy Alternatives.

———. 1989a. Interview with Peter Taylor, 14 April.

———. 1989b. Interview with Peter Taylor by telephone, 11 June.

Picardi, A. C., and W. W. Seifert. 1976. A tragedy of the commons in the Sahel. *Technology Review* (May): 42–51.

Pickering, A. 1992a. From science as knowledge to science as practice. In A. Pickering, ed., *Science as Practice and Culture,* 1–8. Chicago: University of Chicago Press.

———, ed. 1992b. *Science as Practice and Culture.* Chicago: University of Chicago Press.

———. 1993. The mangle of practice: Agency and emergence in sociology of science. *American Journal of Sociology* 99 (3): 559–89.

———. 1995. *The Mangle of Practice: Time, Agency, and Science.* Chicago: University of Chicago Press.

Pickett, S. T. A., and P. S. White, eds. 1985. *The Ecology of Natural Disturbance and Patch Dynamics.* Orlando, FL: Academic Press.

Pielou, E. C. 1969. *An Introduction to Mathematical Ecology.* New York: Wiley-Interscience.

———. 1975. *Ecological Diversity.* New York: Wiley.

Pimm, S. L. 1979. Complexity and stability: Another look at MacArthur's original hypothesis. *Oikos* 33:351–57.

———. 1984. The complexity and stability of ecosystems. *Nature* 307:321–25.

———. 1991. *The Balance of Nature? Ecological Issues in the Conservation of Species and Communities.* Chicago: University of Chicago Press.

Pomerantz, M. J., W. R. Thomas, and M. E. Gilpin. 1980. Asymmetries in population growth regulated by intraspecific competition: Empirical studies and model tests. *Oecologia* 47:311–22.

Post, W. M., and S. L. Pimm. 1983. Community assembly and food web stability. *Mathematical Biosciences* 64:169–92.

Puccia, C. J., and R. Levins. 1985. *Qualitative Modeling of Complex Systems: An Introduction to Loop Analysis and Time Averaging.* Cambridge, MA: Harvard University Press.

Pugh, A. 1973. *Dynamo II User's Manual.* Cambridge, MA: MIT Press.

Radical Science Editorial Collective. 1977. Editorial. *Radical Science Journal* 5:3–7.

Redclift, M., and T. Benton, eds. 1994. *Social Theory and the Global Environment.* London: Routledge.

Reddy, M. 1993. The conduit metaphor: A case of frame conflict in our language about language. In A. Ortony, ed., *Metaphor and Thought,* 164–201. Cambridge: Cambridge University Press.

Resilience Alliance. n.d. http://www.resalliance.org (viewed 10/2/03).

Ribot, J. 1999. Decentralization, participation and accountability in Sahelian forestry: Legal instruments of political-administrative control. *Africa* 69 (1): 23–65.

Roberts, A. 1974. The stability of a feasible random ecosystem. *Nature* 251:607–8.

———. 1979. *The Self-Managing Environment.* London: Allison & Busby.

Roberts, A., and K. Tregonning. 1981. The robustness of natural systems. *Nature* 288:265–66.

Robinson, J. V., and J. E. Dickerson. 1984. Testing the invulnerability of laboratory island communities to invasion. *Oecologia* 61:169–74.

———. 1987. Does invasion sequence affect community structure? *Ecology* 68:587–95.

Robinson, J. V., and M. A. Edgemon. 1988. An experimental evaluation of the effect of invasion history on community structure. *Ecology* 69:1410–17.

Robinson, J. V., and W. D. Valentine. 1979. The concepts of elasticity, invulnerability and invadability. *Journal of Theoretical Biology* 81:91–104.

Robinson, S. 1984. The art of the possible. *Radical Science Journal* 15:122–48.

Roe, E. 1998. *Taking Complexity Seriously: Policy Analysis, Triangulation, and Sustainable Development.* Boston: Kluwer Academic Publishers.

Rose, S., ed. 1982. *Against Biological Determinism: The Dialectics of Biology Group.* London: Allison & Busby.

Rosenberg, C. 1988. Wood or trees?: Ideas and actors in the history of science. *Isis* 79:565–70.

Rosenzweig, M. L. 1973. Evolution of the predator isocline. *Evolution* 27:28–94.

Roughgarden, J. 1977. Basic ideas in ecology. *Science* 196:51.

———. 1983. Competition and theory in community ecology. *American Naturalist* 122:583–601.

Rudwick, M. 1976. The emergence of visual language for geological science, 1760–1840. *History of Science* 14:149–95.

Ruse, M., and P. J. Taylor. 1991. Pictorial representation in biology. Special issue, *Biology & Philosophy* 6:125–294.

Saeed, K. 1982. Public policy and rural poverty: A system dynamics analysis of a social change effort in Pakistan. *Technological Forecasting and Social Change*
21:325–49.

Saunders, P. T. 1978. Population dynamics and the length of food chains. *Nature*
272:189.

Saunders, P. T., and H. J. Bazin. 1975. Stability of complex ecosystems. *Nature*
256:120–21.

Schaffer, W. M. 1981. Ecological abstraction: The consequences of reduced dimen-
sionality in ecological models. *Ecological Monographs* 51:383–401.

Schiebinger, L., ed. 2003. Feminism inside the sciences. *Signs* 28 (3): 859–972.

Science for the People, ed. 1977. *Biology as a Social Weapon*. Minneapolis: Burgess.

Sclove, R. 1995. *Democracy and Technology*. New York: Guilford.

Scott, H. 1933. Science vs. Chaos! In *Introduction to Technocracy*, 7–22. New York:
Technocracy Inc.

———. 1938. *The Mystery of Money*. New York: Technocracy, Inc.

Scott, J. C. 1976. *The Moral Economy of the Peasant: Rebellion and Subsistence in
Southeast Asia*. New Haven, CT: Yale University Press.

Scott, P., E. Richards, and B. Martin. 1990. Captives of controversy: The myth of the
neutral social researcher in contemporary scientific controversies. *Science,
Technology, & Human Values* 15 (4): 474–94.

Segal, H. 1985. *Technological Utopianism in American Culture*. Chicago: University of
Chicago Press.

Seifert, W. W., and K. N. Corones, eds. 1970. *Project Bosporus: Boston Port Utilization
Study*. Cambridge, MA: MIT Press.

Seifert, W. W., and N. Kamrany. 1974. *A Framework for Evaluating Long-Term
Strategies for the Development of the Sahel-Sudan Region*. Cambridge, MA: MIT,
Center for Policy Alternatives.

Selener, D. 1997. *Participatory Action Research and Social Change*. Ithaca, NY:
Cornell University Press.

Sewell, W. H. 1992. A theory of structure: Duality, agency and transformation.
American Journal of Sociology 98:1–29.

Shapin, S. 1982. History of science and its sociological reconstructions. *History of
Science* 20:157–211.

Sheets, H., and R. Morris. 1974. *Disaster in the Desert*. Washington, DC: Carnegie
Endowment for International Peace.

Siljak, D. D. 1975. When is a complex ecosystem stable? *Mathematical Biosciences*
25:25–50.

———. 1978. *Large-Scale Dynamical Systems*. Amsterdam: North-Holland Publishers.

Simberloff, D. 1980. A succession of paradigms in ecology: Essentialism to materialism
to probabilism. *Synthese* 43:3–29.

———. 1982. The status of competition theory in ecology. *Annales Zoologici Fennici*
19:241–53.

Simon, H. 1969. *The Sciences of the Artificial.* Cambridge, MA: MIT Press.

Sinding, S. W., J. A. Ross, and A. G. Rosenfield. 1994. Seeking common ground: Unmet need and demographic goals. *International Family Planning Perspectives* 20 (1): 23–27.

Sismondo, S. 1993a. Response to Knorr Cetina. *Social Studies of Science* 23 (3): 563–69.

———. 1993b. Some social constructions. *Social Studies of Science* 23 (3): 515–53.

Smith, B. H. 1997. Microdynamics of incommensurability: Philosophy of science meets science studies. In B. H. Smith and A. Plotinsky, eds., *Mathematics, Science, and Post-classical Theory,* 243–66. Durham, NC: Duke University Press.

Spencer, H. G., and R. W. Marks. 1988. The maintenance of single-locus polymorphism. I. Numerical studies of a viability selection model. *Genetics* 120:605–13.

Stanfield, B., ed. 1997. *The Art of Focused Conversation.* Toronto: Canadian Institute of Cultural Affairs.

———. 2002. *The Workshop Book: From Individual Creativity to Group Action.* Toronto, Canadian Institute of Cultural Affairs.

Stepan, N. L. 1986. Race and gender: The role of analogy in science. *Isis* 77:261–77.

Sterman, J. 1987. Testing behavioral simulation models by direct experiment. *Management Science* 33:1572–92.

Strauss, S. 1991. Indirect effects in community ecology: Their definition, study and importance. *Trends in Ecology and Evolution* 6 (7): 206–10.

Strong, D. R., D. Simberloff, L. G. Abele, and A. B. Thistle, eds. 1984. *Ecological Communities: Conceptual Issues and the Evidence.* Princeton, NJ: Princeton University Press.

Study of Critical Environmental Problems. 1970. *Man's impact on the global environment.* Cambridge, MA: MIT Press.

Sutter, J. 1977. Sahelian pastoralists: Underdevelopment, desertification and famine. *Annual Review of Anthropology* 6:457–78.

———. 1979. *West African Pastoral Production Systems.* Ann Arbor, MI: Center for Research on Economic Development.

———. 1983. *Cattle and Inequality: A Study of Herd Size Differences among Fulani Pastoralists in Northern Senegal.* Berkeley: Institute for International Studies.

Swift, J. 1977. Sahelian pastoralists: Underdevelopment, desertification and famine. *Annual Review of Anthropology* 6:457–78.

———. 1979. *West African Pastoral Production Systems.* Ann Arbor, MI: Center for Research on Economic Development.

Taylor, P. J. 1979. *The Kerang Farm Model.* Melbourne: Institute of Applied Economic and Social Research, Melbourne, Australia.

———. 1985. Construction and turnover of multispecies communities: A critique of approaches to ecological complexity. Ph.D. dissertation, Harvard University, Cambridge, MA.

———. 1986. Dialectical biology as political practice: An essay review of R. Levins and R. Lewontin, *The Dialectical Biologist. Radical Science* 20:81–111 (Also in L. Levidow, ed., *Science as Politics.* London: Free Association Books.)

———. 1988a. Consistent scaling and parameter choice for linear and Generalized Lotka-Volterra dynamics. *Journal of Theoretical Biology* 135:543–68.

———. 1988b. The construction and turnover of complex community models having Generalized Lotka-Volterra dynamics. *Journal of Theoretical Biology* 135:569–88.

———. 1988c. Technocratic optimism, H. T. Odum, and the partial transformation of ecological metaphor after World War II. *Journal of the History of Biology* 21 (2): 213–44.

———. 1989a. Developmental versus morphological approaches to modeling ecological complexity. *Oikos* 55:434–36.

———. 1989b. Revising models and generating theory. *Oikos* 54:121–26.

———. 1990. Mapping ecologists' ecologies of knowledge. *Philosophy of Science Association* 2:95–109.

———. 1992a. Community. In E. F. Keller and E. Lloyd, eds., *Keywords in Evolutionary Biology,* 52–60. Cambridge, MA: Harvard University Press.

———. 1992b. Re/constructing socio-ecologies: System dynamics modeling of nomadic pastoralists in sub-Saharan Africa. In A. Clarke and J. Fujimura, eds., *The Right Tools for the Job: At Work in Twentieth-Century Life Sciences,* 115–48. Princeton, NJ: Princeton University Press.

———. 1993. What's (not) in the mind of scientific agents? Implicit psychological models and social theory in the social studies of science. Paper presented to Society for Social Studies of Science, West Lafayette, Indiana. http://www.faculty.umb.edu/peter_taylor/4s93.pdf (viewed 1/8/01).

———. 1994a. Notes on causality and explanation in the social sciences. http://www.faculty.umb.edu/peter_taylor/socsciexpl.html (viewed 1/8/01).

———. 1994b. Shifting frames: From divided to distributed psychologies of scientific agents. *Proceedings of the Philosophy of Science Association* 2:304–10.

———. 1995a. Building on construction: An exploration of heterogeneous constructionism, using an analogy from psychology and a sketch from socio-economic modeling. *Perspectives on Science* 3 (1): 66–98.

———. 1995b. Co-construction and process: A response to Sismondo's classification of constructivisms. *Social Studies of Science* 25:348–59.

———. 1996a. Notes on metaphor readings, made with a view to stimulating discussion and clarification. http://www.faculty.umb.edu/peter_taylor/metaphor.html (viewed 12/20/00).

———. 1996b. Science and Social Theory (theme: structure and agency): Syllabus for STS662, Cornell University, Spring semester. http://www.faculty.umb.edu/peter_taylor/662a-96.html (viewed 12/20/00).

———. 1997a. Afterword: Shifting positions for knowing and intervening in the cultural politics of the life sciences. In P. J. Taylor, S. E. Halfon, and P. N. Edwards, eds., *Changing Life: Genomes, Ecologies, Bodies, Commodities,* 202–24. Minneapolis: University of Minnesota Press.

———. 1997b. Appearances nonwithstanding, we are all doing something like political ecology. *Social Epistemology* 11 (1): 111–27.

————. 1997c. How do we know we have global environmental problems? Undifferentiated science-politics and its potential reconstruction. In P. J. Taylor, S. E. Halfon, and P. E. Edwards, eds., *Changing Life: Genomes, Ecologies, Bodies, Commodities,* 149–74. Minneapolis: University of Minnesota Press.

————. 1997d. Making connections and respecting differences: Reconciling schemas for learning and group process. *Connexions (Newsletter of International Society for Exploring Teaching Alternatives)* (March & July); reproduced as http://www.faculty.umb.edu/peter_taylor/connexions.html (viewed 11/1/03).

————. 1998a. How can we make complexity facilitate social change? Session held at the International Society for Exploring Teaching Alternatives, Cocoa Beach, FL, October. http://www.faculty.umb.edu/peter_taylor/iseta98.html (viewed 12/20/00).

————. 1998b. How does the commons become tragic? Simple models as complex socio-political constructions. *Science as Culture* 7 (4): 449–64.

————. 1998c. Natural selection: A heavy hand in biological and social thought. *Science as Culture* 7 (1): 5–32.

————. 1999a. Basic propositions of the workshop process. http://www.faculty. umb.edu/peter_taylor/ICApropositions.html (viewed 12/12/99).

————. 1999b. Mapping the complexity of social-natural processes: Cases from Mexico and Africa. In F. Fischer and M. Hajer, eds., *Living with Nature: Environmental Discourse as Cultural Critique,* 121–34. Oxford: Oxford University Press.

————. 1999c. What can agents do? Engaging with complexities of the post-Hardin commons. In L. Freese, ed., *Advances in Human Ecology,* vol. 8, 125–56. Greenwich, CT: JAI Press.

————. 2000a. Assessing biodiversity and ecological stability. *Science* 290:51.

————. 2000b. Knowledge-making, social agency, and complexity in environmental analysis. Paper presented to the Environmental Social Science Conference, University of Tampere, Finland, May.

————. 2000c. Socio-ecological webs and sites of sociality: Levins' strategy of model building revisited. *Biology & Philosophy* 15 (2): 197–210.

————. 2001a. Distributed agency within intersecting ecological, social, and scientific processes. In S. Oyama, P. Griffiths, and R. Gray, eds., *Cycles of Contingency: Developmental Systems and Evolution,* 313–32. Cambridge, MA: MIT Press.

————. 2001b. From critical thinking to reflective practice, especially about environment, science, and society. http://www.faculty.umb.edu/peter_taylor/portfolio01-TOC.html#Statement (viewed 7/15/03).

————. 2001c. From natural selection to natural construction to disciplining unruly complexity: The challenge of integrating ecological dynamics into evolutionary theory. In R. Singh, K. Krimbas, D. Paul, and J. Beatty, eds., *Thinking about Evolution: Historical, Philosophical and Political Perspectives,* 377–93. Cambridge: Cambridge University Press.

————. 2001d. How do we know there is a population-environment problem? http://www.faculty.umb.edu/peter_taylor/popdialogue.html (viewed 7/13/01).

————. 2002a. Non-standard lessons from the "tragedy of the commons." In M. Maniates, eds., *Empowering Knowledge: A Primer for Teachers and Students of Global Environmental Politics*, 87–105. New York: Rowman & Littlefield.

————. 2002b. We know more than we are, at first, prepared to acknowledge: Journeying to develop critical thinking. http://www.faculty.umb.edu/peter_taylor/journey.html (viewed 11/1/03).

————. 2004. "Whose trees/interpretations are these?" Bridging the divide between subjects and outsider-researchers. In R. Eglash, J. Croissant, G. DiChiro, and R. Fouché, eds., *Appropriating Technology: Vernacular Science and Social Power*, 305–12. Minneapolis: University of Minnesota Press.

Taylor, P. J., and A. S. Blum 1991a. Ecosystems as circuits: Diagrams and the limits of physical analogies. *Biology & Philosophy* 6:275–94.

————. 1991b. Pictorial representation in biology. *Biology & Philosophy* 6:125–34.

Taylor, P. J., and F. H. Buttel. 1992. How do we know we have global environmental problems? Science and the globalization of environmental discourse. *Geoforum* 23 (3): 405–16.

Taylor, P. J., and R. García-Barrios. 1995. The social analysis of ecological change: From systems to intersecting processes. *Social Science Information* 34 (1): 5–30.

————. 1997. The dynamics and rhetorics of socio-environmental change: Critical perspectives on the limits of neo-Malthusian environmentalism. In L. Freese, ed., *Advances in Human Ecology*, vol. 6, 257–92. Greenwich, CT: JAI Press.

Taylor, P. J., and Y. Haila. 1989. Mapping workshops for teaching ecology. *Bulletin of the Ecological Society of America* 70 (2): 123–25.

————. 2001. Situatedness and problematic boundaries: Conceptualizing life's complex ecological context. *Biology & Philosophy* 16 (4): 521–32.

Taylor, P. J., and W. M. Post. 1985. *A description with some applications of MSNUCY, a computer model combining interspecific interactions with nutrient cycling.* Environmental Sciences Division publication 2419. Oak Ridge, TN: Oak Ridge National Laboratory.

Thompson, C. 2002. When elephants stand for competing philosophies of nature: Amboseli National Park, Kenya. In J. Law and A. Mol, eds., *Complexities: Social Studies of Knowledge Practices,* 166–90. Durham, NC: Duke University Press.

Tilman, D. 1999. The ecological consequences of changes in biodiversity: A search for general principles. *Ecology* 80:1455–74.

Tilman, D., and S. Pacala. 1993. The maintenance of species richness in plant communities. In R. Ricklefs and D. Schluter, eds., *Species Diversity in Ecological Communities,* 13–25. Chicago: University of Chicago Press.

Traweek, S. 1994. Worldly diffractions: Feminist and cultural studies of science, technology, and medicine. Paper presented at the annual meeting of the American Sociological Association, Los Angeles, August.

Tregonning, K., and A. P. Roberts. 1979. Complex systems which evolve towards homeostasis. *Nature* 281:563–64.

Tucker, R. C., ed. 1978. *The Marx-Engels Reader*. New York: W. W. Norton.

Tufte, E. R. 1983. *The Visual Display of Quantitative Information*. Cheshire, CT: Graphics Press.

Turner, M. D. 1993. Overstocking the range: A critical analysis of the environmental science of Sahelian pastoralism. *Economic Geography* 69 (4): 402–21.

———. 1999. The role of social networks, indefinite boundaries and political bargaining in maintaining the ecological and economic resiliency of the transhumance systems of Sudano-Sahelian West Africa. In M. Niamir-Fuller, ed., *Managing Mobility in African Rangelands: The Legitimization of Transhumance*, 97–123. London: Intermediate Technology Publications.

———. 2003. Methodological reflections on the use of remote sensing and Geographic Information Science in human ecological research. *Human Ecology* 31 (2): 255–79.

Turner, M. D., and P. J. Taylor, eds. 2003. Critical reflections on the use of remote sensing and GIS technologies in human ecological research. *Human Ecology* 31 (2): 179–306.

Turner, S. 1994. *The Social Theory of Practices: Tradition, Tacit Knowledge, and Presuppositions*. Chicago: University of Chicago Press.

Ulanowicz, R. E. 1986. *Growth and Development*. New York: Springer Verlag.

Underwood, A. J. 1997. *Experiments in Ecology: Their Logical Design and Interpretation Using Analysis of Variance*. Cambridge: Cambridge University Press.

United Nations. 1973. Final report on the meeting of the Sudano-Sahelian Mid- and Long-term Programme. New York: United Nations, Special Sahelian Office.

United States Agency for International Development (USAID). 1973. Supporting document: Appendix 2 to United States Congress 1973. Washington, DC: United States Agency for International Development.

———. 1975. *Development Assistance Program, 1976–80, Central and West Africa Region*. Washington, DC: United States Agency for International Development.

———. 1976. Proposal for a long-term comprehensive development program for the Sahel: Report to the U.S. Congress, April 1976. Washington, DC: United States Agency for International Development.

United States Congress. Senate. Committee of the Judiciary. 1973. *Hearing before the Subcommittee to investigate problems connected with refugees and escapees*. 93rd Cong., 25 July. Washington, DC: United States Senate.

Urwin, C. 1984. Power relations and the emergence of language. In J. Henriques, W. Holloway, C. Urwin, C. Venn, and V. Walkerdine, eds., *Changing the Subject: Psychology, Social Regulation and Subjectivity*, 264–322. London: Methuen.

USAID. *See* United States Agency for International Development.

Vandermeer, J. H. 1969. The competitive structure of communities: An experimental approach with protozoa. *Ecology* 50:362–71.

———. 1977. Ecological determinism. In Science for the People, ed., *Biology as a Social Weapon*, 108–22. Minneapolis: Burgess.

———. 1980. Indirect mutualism: Variations on a theme by Stephen Levine. *American Naturalist* 116:441–48.

———. 1981. A further note on community models. *American Naturalist* 117:379–80.

———. 1990. *Elementary Mathematical Ecology*. Malabar, FL: Krieger.

Vayda, A. P., and B. B. Walters. 1999. Against political ecology. *Human Ecology* 27: 167–79.

Vernadsky, V. 1944. Problems of biogeochemistry, II: The fundamental matter-energy difference between the living and inert natural bodies of the biosphere. *Transactions of the Connecticut Academy of Arts and Science* 35:485–517.

Vogt, E. Z. 1960. On the concepts of structure and process in cultural anthropology. *American Anthropology* 62:18–33.

Vonk, R. 1987. Indigenous agroforestry systems: A case study from the CARE-Kenya agroforestry extension program. Parts I & II. Papers delivered to the Social Forestry Program at the University of California, Berkeley, 21–22 September.

Waddington, C. H., ed. 1969. *Towards a Theoretical Biology*. Edinburgh: Edinburgh University Press.

Walters, C. 1986. *Adaptive Management of Renewable Resources*. New York: Macmillan.

———. 1997. Challenges in adaptive management of riparian and coastal ecosystems. *Conservation Ecology* 1 (2): 1 [http://www.consecol.org/vol1/iss2/art1].

Weiher, E., and P. Keddy, ed. 1999. *Ecological Assembly Rules: Perspectives, Advances, Retreats*. Cambridge: Cambridge University Press.

Weinberg, A. 1961. Impact of large-scale science on the United States. *Science* 134:161–64.

Weissglass, J. 1990. Constructivist listening for empowerment and change. *Educational Forum* 54 (4): 351–70.

Werskey, G. 1988 [1978]. *The Visible College: A Collective Biography of British Scientists and Socialists of the 1930s*. London: Free Association Books.

West Nipissing Economic Development Corporation. 1993. Vision 20/20: Shaping our futures together, executive summary. April 1993.

———. 1999. Vision 2000 Plus, executive summary. June.

White, M. 1998. Notes on narrative metaphor and narrative therapy. In C. White and D. Denborough, eds., *Introducing Narrative Therapy: A Collection of Practice-Based Writings,* 225–27. Adelaide: Dulwich Centre Publications.

White, M., and D. Epston. 1990. *Narrative Means to Therapeutic Ends*. New York: W. W. Norton.

Whole Earth Group. 1974. *Uncle Afrely's Earth Guide: A Handbook*. Clayton, Australia: Monash University Student Association.

Wiener, N. 1948a. *Cybernetics*. Cambridge, MA: MIT Press.

———. 1948b. Time, communication, and the nervous system. *Annals of the New York Academy of Sciences* 50:197–219.

Williams, F. M. 1972. Mathematics of microbial populations, with an emphasis on open systems. *Transactions of the Connecticut Academy of Arts and Sciences* 44:397–426.

Williams, R. 1980. Ideas of nature. In *Problems in materialism and culture,* 67–85. London: Verso.

———. 1983. *The Year 2000.* New York: Pantheon.

———. 1985. *Loyalties.* London: Chatto & Windus.

———. 1990. *People of the Black Mountains: The Beginning.* London: Palladin.

———. 1992. *People of the Black Mountains: The Eggs of the Eagle.* London: Palladin.

Wimsatt, W. C. 1987. False models as a means to truer theories. In M. Nitecki and A. Hoffman, eds., *Neutral Models in Biology,* 23–55. New York: Oxford University Press.

———. 2001. Generative entrenchment and the developmental systems approach to evolutionary processes. In S. Oyama, P. E. Griffiths, and R. D. Gray, eds., *Cycles of Contingency: Developmental Systems and Evolution,* 219–37. Cambridge, MA: MIT Press.

Winstanley, D. 1973. Recent rainfall trends in Africa, the Middle East, and India. *Nature* 243:464–65.

Wise, N. 1988. Mediating machines. *Science in Context* 2 (1): 77–113.

Wolf, E. 1982. Afterword. In *Europe and the People without History,* 385–91. Berkeley: University of California Press.

Wondolleck, J. M., and S. L. Yaffee. 2000. *Making Collaboration Work: Lessons from Innovation in Natural Resource Management.* Washington, DC: Island Press.

Woodhouse, E. J. 1991. The turn toward society? Social reconstruction of science. *Science, Technology & Human Values* 16 (3): 390–404.

Woolgar, S. 1981. Interests and explanation in the social study of science. *Social Studies of Science* 11:365–94.

———, ed. 1988. *Knowledge and Reflexivity: New Frontiers in the Sociology of Knowledge.* London: Sage.

Wootton, J. T. 1994. The nature and consequences of indirect effects in ecological communities. *Annual Review of Ecology and Systematics* 25:443–66.

Worthington, E. B. 1975. *The Evolution of the IBP.* Cambridge: Cambridge University Press.

Yearley, S. 1991. *The Green Case: A Sociology of Environmental Issues, Arguments and Politics.* London: HarperCollins Academic.

Young, R. M. 1985. *Darwin's Metaphor: Nature's Place in Victorian Culture.* Cambridge: Cambridge University Press.